KINSHIP: BELONGING IN A WORLD OF RELATIONS
VOLUME 4: PERSONS

D1738128

VOL. 04

PERSONS

Edited by

Gavin Van Horn, Robin Wall Kimmerer, John Hausdoerffer

Center for Humans and Nature Press

Center for Humans and Nature Press, Libertyville 60030
© 2021 by Center for Humans and Nature

All rights reserved. No part of this book may be used or reproduced in any manner whatsoever without written permission, except in the case of brief quotations in critical articles and reviews.

For more information, contact the Center for Humans and Nature Press, 17660 West Casey Road, Libertyville, Illinois 60048.
Printed in the United States of America.

Cover and slipcase design: LimeRed, https://limered.io

ISBN-13: 978-1-7368625-0-6 (paper)
ISBN-13: 978-1-7368625-1-3 (paper)
ISBN-13: 978-1-7368625-2-0 (paper)
ISBN-13: 978-1-7368625-3-7 (paper)
ISBN-13: 978-1-7368625-4-4 (paper)
ISBN-13: 978-1-7368625-5-1 (set/paper)

Names: Van Horn, Gavin, editor | Kimmerer, Robin Wall, editor | Hausdoerffer, John, editor
Title: Kinship: belonging in a world of relations, vol. 4, persons / edited by Gavin Van Horn, Robin Wall Kimmerer, and John Hausdoerffer

Description: First edition. | Libertyville, IL: Center for Humans and Nature Press, 2021 | Identifiers: LCNN 2021909501 | ISBN 9781736862537 (paper)

Copyright and permission acknowledgments appear on page 144.

Center for Humans and Nature Press
17660 West Casey Road, Libertyville, Illinois 60048

www.humansandnature.org

Printed by Graphic Arts Studio, Inc. on Rolland Opaque paper. This paper contains 30% post-consumer fiber, is manufactured using renewable energy - Biogas and is elemental chlorine free. It is Forest Stewardship Council® and Rainforest Alliance certified.

CONTENTS

KINNING: INTRODUCING THE KINSHIP SERIES

Gavin Van Horn

The lines hung lightly suspended in midair. Twinkling, illuminated from within. Then vanished. I inclined my head. The lines reappeared, seeming to materialize out of emptiness. These weren't merely lines; they were radial strands precisely strung from a central axis, intricately woven. The glow from a nearby streetlamp caught within them, revealing a sacred geometry. I leaned closer. With a slight tilt of my face—altering the angle between my eyes and the spider's evening project—the lines alternately disappeared or disclosed themselves. Their creator, deftly putting the finishing touches on this work of body art, was smaller than my thumbnail. Her handiwork sparkled with its own radiance. For a moment I felt envious—then grateful. The craft on display demonstrated skills of which I was completely and utterly incapable. I drew in closer to get a better look at the stitching, which would likely hang for the evening before being pulled apart by a strong morning wind.

Kinship: Belonging in a World of Relations can be described as a series of books, five volumes that group different essays and poems according to the scale of their subject matter, from the composition of the cosmos to the gestures of the everyday: Planet, Place, Partners, Persons, Practice. But these books could just as easily be described as a web, a meshwork whose strands gather, crisscross, and link together a vast variety of subjects and experiences. Each book reaches beyond its pages, spinning silk filaments through the others; turn your head at the right angle and an intricate web appears—functional, sensorial, and artful.

The essays and poems you hold in your hands comprise lines, strands of ink, patterns on paper. In your imagination, these words may come to life, recalling and revealing shared relations with our fellow Earthlings—our kinfolk—who come in all shapes and sizes, from the bacterium swimming in your belly or lying on the tip of your tongue to the vibrant collective breath that sweeps across your face and into your lungs. Worth thinking about—and perhaps *thanking* about—are the shared threads between kinfolk, especially plantfolk, that make this breath exchange possible. Your life, my life, all of our lives depend on the quality of relations between us—the air we breathe, the water we drink, the food we eat and the food we become—within an exuberant, life-generating planetary tangle capable of nurturing intelligences that can spin webs and words.

Kinning

The words in this *Kinship* web gesture beautifully toward the relations—vital, wild processes—that are always present yet not always visible. Because these relations may be difficult to apprehend, it may seem as though the world is merely a collection of inert objects, full of nouns. You are you. I am me. That bird at the feeder is, sadly, referred to as "it." That river underneath the bridge and that mountain on the horizon are designated "natural resources." Some of us have rights, legal standing, personhood. Some of us—depending on which nation-state we happen to dwell in—don't.

Nouns have utility, yet they can mislead, perhaps even reifying the idea that the world is composed of things—some small, some large, some shiny, some dull, some with wings, some with legs, some with leaves, some with fur. This language-induced reduction would suggest everything is mere matter, a gathering of atoms, in more or less complicated geometries. Note, however, even in that last sentence, an interesting gerund slipped into the mix. What are atoms but a gathering of *relations*? What is a gathering of relations if not *relatives relating*? Just as when I tilted my head and the spiderweb "appeared," due to a slight change of perspective, it is possible with a shift of perspective to see the threads connecting worlds, all the relations that make us kin. The point here is one that comes up repeatedly in *Kinship*: Earth—and everything within it, including all that creates what we call earth—is a verb. All is in motion; all is relating.

The English language is noun dominant, and in comparison to many Indigenous languages, the animacy and agency of other beings and processes often receive less emphasis. Though obviously still relying on English, these *Kinship* volumes—because of their subject matter—challenge this object obsession at its core. *Kinship* can be considered a noun, of course, a state of being—whether this is couched in terms of biological genetics; family, clan, or species affiliation; shared and storied relations and memories that inhere in people and places; or more metaphorical imaginings that unite us to faith traditions, cultures, countries, or the planet. But the voices in these volumes point us toward an alternative perspective: kinship *as a verb*.

Perhaps this kinship-in-action should be called kinning. Humans are born kin, in any number of ways. But the words in this *Kinship* anthology collectively express something more than birthright claims: they point toward how it is possible to *become* kin. In this understanding, being kin is not so much a given as it is an intentional process. Kinning does not depend upon genetic codes. Rather, it is cultivated by humans, as one expression of life among many, many, many others, and it revolves around an ethical question: how to rightly relate? We are kinning as we (re)connect our bodies, minds, and spirits within a world that is not merely a collection of objects but "a communion of subjects," as Thomas Berry put it.[1] The essays and poetry in these volumes, at different scales and in different geographies, show possibilities for becoming better kin—more receptive to the languages of others, especially nonhuman others, and better listeners to their stories, which reach out to us through place and time. This vibrant world, as well as these volumes, offers invitations for kinning—practices of belonging with and amid our fellow earthly kin.

Three Threads in the Web

Three conceptual threads came together to inform and inspire the creation of the multiple *Kinship* volumes. I'd like to briefly mention them here, as readers may want to be alert to them throughout various essays in the books.[2]

The first thread is a cosmovision—and, increasingly, legal recognition—that acknowledges and understands nonhumans, including entire watersheds, forests, or mountains, as persons. In the West, many of us have inherited a settler-colonial worldview that uplifts the human individual—often implicitly or explicitly identified with whiteness and maleness—as the locus of meaning and center of importance while reducing nature to resources, property, or fungible commodities. Bootstrap economics and the literary hero's journey reinforce such thinking—the lone figure encountering and overcoming obstacles, conquering beasts, and emerging victorious above the fracas. From this vantage, there are human persons (and now corporate "persons"), and there is everything else.

The religious studies scholar Graham Harvey, in his wide-ranging study of animistic and neo-animistic cultures and movements, upends such notions. Harvey observes that from an animistic perspective, "the world is full of persons, only some of whom are human." Persons, he goes on to write, are not equated solely with human beings in many cultures. Rather, the term serves as a broader umbrella for those beings who are perceived as displaying agency (and this encompasses landscapes, rocks, and bodies of water, in addition to plants and nonhuman animals):

> Persons are beings, rather than objects, who are animated and social towards others (even if they are not always sociable).

> Animism *may* involve learning how to recognize who is a person and what is not—because it is not always obvious and not all animists agree that everything that exists is alive or personal. However, animism is more accurately understood as being concerned with learning how to be a good person in respectful relationships with other persons.[3]

When I first read Harvey fifteen years ago or so, I didn't know how or if this type of cosmology could ever make its way into mainstream Western consciousness. Then, in March 2017, the Whanganui River (Te Awa Tupua), the third-largest river in Aotearoa New Zealand, grabbed international headlines. The Whanganui officially gained legal status as a living entity with the same rights of personhood as a human being. More than a change in legal nomenclature, this reclassification of the river stands as a significant bicultural effort to bring disparate systems of law and care together among New Zealanders of European descent and the native Māori population (Te Āti Haunui-a-Pāpārangi).[4]

The designation of the Whanganui River is one instance in a growing number of cases in which legal personhood is being granted to nonhuman entities. An overlapping set of localized and national governmental precedents, many of which involve personhood language, for example, began to gain traction in 2006 by focusing on the "rights of nature."[5] Ecuador and Bolivia both included rights-of-nature clauses in their national constitutions in 2008 and 2010, respectively. In Colombia, courts ruled in favor of personhood for the Amazon and Atrato Rivers. In 2016, the Ho-Chunk Nation in Wisconsin amended their tribal constitution to include rights-of-nature language: "Ecosystems, natural communities, and species within the Ho-Chunk Nation territory possess inherent, fundamental, and inalienable rights to naturally exist, flourish, regenerate, and evolve." In 2017, the Ponca Nation in Oklahoma recognized rights of nature as statutory law to combat fracking. Australia, India, and Nepal have also taken steps toward establishing rights of nature. In 2019, the Yurok Tribal Council passed a resolution that declared the personhood of the Klamath River in the Pacific Northwest. Such landmark legal and legislative actions represent efforts to give "voice" to other-than-human beings, ensuring their inherent rights to exist and flourish. Gerard Albert, the lead Maori negotiator on behalf of the Whanganui *iwi* (tribe), summed up this sense of responsibility well: "We can trace our genealogy to the origins of the universe. And therefore, rather than us being masters of the natural world, we are part of it. We want to live like that as our starting point. And that is not an anti-development, or anti-economic use of the river but to begin with the view that it is a living being, and then consider its future from that central belief."[6]

Recognition of kinship has many overlaps with these attempts to recognize personhood. The commonality lies in a respect for the agency of other beings and concerted efforts to treat them with dignity and even deference. This brings us to the second thread. *Kincentric ecology*, a phrase coined by ethnobotanist Enrique Salmón, provides a helpful guide for understanding kinship: an intertwining of the social, mythological, and practical. Salmón asserts that "life in any environment is viable only when humans view their surroundings as kin; that their mutual roles are essential for their survival."[7] This perspective stands in marked contrast to the familiar, if not predominant, human chauvinism toward other species in so many national sociopolitical systems. From a kinship perspective, the landscapes of which humans are a part—including rocks, rivers, oceans, prominent geographic features, and other nonhuman plant and animal persons—provide a shared sense of place and require appropriate human care and respect.

This kinship is deep and wide—and dwells within the human body. In the past century and a half, evolutionary and ecological sciences have brought additional insights to bear on what it means to be human. In only the past few decades, evolutionary models are being transformed by research into symbiotic mergers at the cellular level, horizontal gene transfer, and seemingly chimeric creatures that rely on cooperative relationships between species from entirely different "kingdoms" of life. Kinship, it would seem, is key to understanding biotic and abiotic entanglement. A kincentric ecology emerges from cultures that recognize the importance of humans in maintaining right relations in particular landscapes. Far from presupposing that humans are a degrading force, sullying whatever we might touch, a kincentric ecology expresses the view that humans can actually play keystone roles in our landscapes, creating mutual flourishing. In other words, human beings are not merely kin by biological relation, but it is entirely possible that human communities and cultures can be good kin, salutary ecological collaborators alongside and with our nonhuman family members.

The third thread that inspired this *Kinship* series, I'm happy to say, comes from coeditor Robin Wall Kimmerer. Robin's work draws from scientific training and Indigenous knowledge in complementary ways. In *Braiding Sweetgrass: Indigenous Wisdom, Scientific Knowledge and the Teachings of Plants*, she explores her own history of loss and recovery as a member of the Citizen Potawatomi Nation and describes how Indigenous perspectives can transform engagement with a living world. Perhaps nowhere is this clearer than when she contrasts the "grammar of animacy" embedded in the Potawatomi language to conventional English and its objectifying pronouns. She makes a convincing case that an ethical revolution might depend on a language revolution. Finding ways to properly and respectfully acknowledge *ki* (the pronoun Robin proposes for our other-than-human kin) is a good place to begin.

You will recognize these three threads—nonhuman personhood, humans as relational participants in local ecologies, and the care expressed when addressing and engaging with our kinfolk through language—winding their way through all the volumes of *Kinship*.

Five Scales of Kinning

With all the amazing contributors gathered in this *Kinship* web, it might help the reader to know what we as editors were asking of them. The following are the questions we posed to our contributors for each *Kinship* volume, which feature the ways kinship can be understood at different scales: from deep time cosmic and evolutionary relationships; to community watersheds, landscapes, and bioregions; to interspecies engagements and mythological perceptions; to biological and symbolic understandings of human interbeing; to which kinds of practices are appropriate for making and becoming kin:

> *Volume 1: Planet*—With every breath, every sip of water, every meal, we are reminded that our lives are inseparable from the life of the world—and the cosmos—in ways both material and spiritual. What are the sources of our deepest evolutionary and planetary connections, and of our profound longing for kinship?
>
> *Volume 2: Place*—Given the place-based circumstances of human evolution and culture, global consciousness may be too broad a scale of care for us. To what extent does crafting a deeper connection with Earth's bioregions reinvigorate a sense of kinship with the place-based beings, systems, and communities that mutually shape one another?
>
> *Volume 3: Partners*—How do cultural traditions, narratives, and mythologies shape the ways we relate, or not, to other beings as kin? How do relations between and among different species foster a sense of responsibility and belonging in us?
>
> *Volume 4: Persons*—Kinship spans the cosmos, but it is perhaps most life changing when experienced directly and personally. Which experiences expand our understanding of being human in relation to other-than-human beings? How can

we respectfully engage a world full of human and nonhuman persons?

Volume 5: Practice—From the perspective of kinship as a recognition of nonhuman personhood, of kincentric ethics, and of *kinship* as a verb involving active and ongoing participation, how are we to live? What are the practical, everyday, and lifelong ways we *become* kin?

We invited our contributors because of their experiences, their expertise, their diverse backgrounds and geographical locations, and because of the way they've made kin with particular species—or some combination of all of these. We also invited our contributors to share their words because of their abilities to tell a good story.

In many of the essays in *Kinship*, there are statistics, references to academic sources, explorations of complicated ideas, and endnotes that may take a reader onward, but we above all wanted readers to hear people's stories. As human beings, we are storytelling animals. We lean a little closer when we overhear someone else say, "Oh my, have I got a good story for you!" For similar reasons, our Paleolithic ancestors likely leaned into the firelight—as it crackled against the cave walls at Lascaux, France, or Sulawesi, Indonesia—watching the aurochs, bison, horses, and deer dance before their eyes, or the warty pig and the babirusa (pig-deer).[8]

As storytelling creatures, when thinking of our personal relationships with the natural world, we may be predisposed to be on the lookout for epiphanies—the holy overwhelm, the big payoff, the road-to-Damascus moment, the final boss battle. But becoming kin, as the various stories in these volumes attest, consists of repeated intimacies, familiar encounters, and daily undoings and transformations that are dependent on visitations and conversations within a smaller circle of place. Awe-filled moments of raw contact with forces that relativize human importance should not be disparaged or discounted. Such experiences may stir deep wells of gratitude. But knowing that humans are relatives, responsible for the lives of others, as others are responsible for ours, remains at the heart of these volumes. If humans are relatives relating, not merely in terms of an abstract genetic code but as intimate familiars, the question then becomes how we as individual persons and communities might better cultivate these relations. How can we uproot the desire to impose our will upon the living worlds around us? How do we become more receptive to nonhuman languages and ways of being?

One step in this direction is the recognition that nature is not a passive object, a text awaiting our interpretation or exegesis, a thing humans approach solely for insights, entertainment, and "resources." The world all of us are part of and participate in is a relational exchange—alive, wildly generative, an ongoing conversation of bodies, desires, conflicts, and collaborations. There is no pinnacle here for humans to sit atop and gaze upon the masses. *Kinship* culminates, in volume 5, with a wide-ranging conversation about practices and ethics that embrace a world of other-than-human persons as worthy of our active care, concern, and respect. At a time when human fidelity to the natural world seems to be fraying, *Kinship* offers stories of solidarity, highlighting the deep interdependence that exists between humans and the more-than-human world. It explores challenging questions, including how communities might fairly and effectively give voice to nonhuman beings and landscapes. And it highlights the cosmologies, mythical narratives, and everyday practices that embrace a world of other-than-human persons as worthy of response and responsibility.

Humans will survive and continue to tell our stories if we learn how to live well with our kin. The voices included in these volumes—these webs of words—and the collective wisdom they express, invite us into this kind of kinning. These are the stories of how to listen to voices other than our own.

Lean a little closer into the firelight. Up on the ceiling of the cave, or near the streetlamp, or between branches in the forest, or in the corner of the room within which you are currently reading, a web may be flashing in a flicker of light.

NOTES

1. Thomas Berry, *Evening Thoughts: Reflecting on Earth as Sacred Community* (San Francisco: Sierra Club, 2006), 149. More on Berry as "geologian" can be found at the website of the Thomas Berry Foundation, http://thomasberry.org/life-and-thought/about-thomas-berry/geologian.
2. In addition to the themes, the three persons mentioned all have essays in the *Kinship* volumes. Graham Harvey, "Academics Are Kin, Too: Transformative Conversations in the Animate World," appears in *Vol. 4: Persons*; Enrique Salmón, "A Heart Rooted in Place: Poetic Dentists and Getting Rained On," is in *Vol. 2: Place*; and Robin Wall Kimmerer, "A Family Reunion near the End of the World," is in *Vol. 1*: Planet.
3. Graham Harvey, *Animism: Respecting the Living World* (New York: Columbia University Press, 2006), xi.
4. See Anna M. Gade, "Managing the Rights of Nature for Te Awa Tupua," *Edge Effects*, September 5, 2019, https://edgeeffects.net/te-awa-tupua/. As my coeditor Robin pointed out to me, "This can be understood, not [that] the river *gained* personhood" but "more that Western institutions came to acknowledge its intrinsic personhood, under the tutelage of the Maori, who have always recognized this inherent nature."
5. Craig M. Kauffman and Pamela L. Martin, "Constructing Rights of Nature Norms in the U.S., Ecuador, and New Zealand," *Global Environmental Politics* 18, no. 4 (2018): 43–62, https://www.mitpressjournals.org/doi/pdf/10.1162/glep_a_00481.
6. E. A. Roy, "New Zealand River Granted Same Legal Rights as Human Being," *The Guardian*, March 16, 2017, https://www.theguardian.com/world/2017/mar/16/new-zealand-river-granted-same-legal-rights-as-human-being.
7. Enrique Salmón, "Kincentric Ecology: Indigenous Perceptions of the Human-Nature Relationship," *Ecological Applications* 10, no. 5 (2000): 1327–32, https://www.researchgate.net/profile/Enrique_Salmon/publication/242186767_Kincentric_Ecology_Indigenous_Perceptions_of_the_HumanNature_Relationship/links/5c34e542a6fdccd6b59c2aa1/Kincentric-Ecology-Indigenous-Perceptions-of-the-HumanNature-Relationship.pdf.
8. Leang Timpuseng is the name of the cave in Sulawesi, Indonesia, whose figurative art has been geochemically dated using recently developed techniques. The results pushed back the timeline on some of the earliest known figurative cave art, and examples of symbolic thinking, to more than 35,000 years ago. "Find early paintings, particularly figurative representations like animals, and you've found evidence for the modern human mind," writes Jo Marchant in "A Journey to the Oldest Cave Paintings in the World," *Smithsonian Magazine*, January–February 2016, https://www.smithsonianmag.com/history/journey-oldest-cave-paintings-world-180957685/. The timeline continues to lengthen; in 2021, from a karst system in Sulawesi, a date of 45,500 years ago was announced for the earliest known representational work of art. See Adam Brumm et al., "Oldest Cave Art Found in Sulawesi," *Science Advances* 7, no. 3 (January 13, 2021): https://advances.sciencemag.org/content/7/3/eabd4648.

STARLINGS, INFINITY, AND THE KITH OF KINSHIP

Lyanda Fern Lynn Haupt

A starling lives in my house. I prefer to think of the circumstance under which she came into my care as a heroic rescue, but to be perfectly honest, it was part theft. I was working on a book about kinship and creativity, explored through the window of Mozart's relationship with a starling he kept for four years. In my research for the project, I'd scoured the academic literature on starlings, interviewed myriad experts, and traveled to Austria, where I haunted the Vienna apartment in which Mozart had lived and composed alongside the bird. But the longer I pondered and scribbled, the more I came to recognize a gap in my understanding of the human-bird relationship central to my project—I didn't know, as Mozart did, what it was like to coexist with a starling in my own household.

Under the Migratory Bird Treaty Act (and according to common ecological sense) it is entirely illegal to remove, harass, touch, or even glance sideways at the nests, eggs, or nestlings of nearly any bird. Starlings are one of the few exceptions. The twenty-some starlings that were introduced to Central Park in the late 1800s have swelled in number to two hundred million and now blanket North America, flourishing on farmland, in suburban lawns, and in urban parks. They are omnivorous, adaptive, and smart. While most of the general public cannot accurately identify starlings, we do know one thing: we aren't supposed to like them. In ornithological and conservation circles, starlings are beyond question the most despised bird on the continent, competing with native cavity nesters such as acorn woodpeckers and bluebirds for prized nest holes. They cause millions of dollars in agricultural damage each year, their great flocks descending to feast upon fields and animal feed. With this track record, the starling is considered not just an introduced species but an invasive one; fish and wildlife departments across the country enjoin us to discourage starlings from nesting, cover their nest holes, destroy their nests, remove their eggs, and even kill both chicks and adult birds in any way we can think up.

One day a friend in the city parks department familiar with my project called to let me know that a starling nest in the eaves of a nearby park bathroom was going to be removed by park workers. I'd been studying that nest and knew that the young had just hatched—here was my chance. It was a harrowing caper, but with my husband as coconspirator, some ill-conceived climbing upon wobbly park-owned garbage cans, a share of bloody scrapes and ugly bruises, and a shocking amount of foul language, I managed to swoop in and scoop up a nestling before she was swept into the city trash bin. I was aware that while we are allowed to maim or murder starlings with legal impunity, it is decidedly not legal to lovingly nurture one in our living rooms. I have a background in avian rehab and knew what I was doing in terms of raising a chick. Still—I was about to become a minor criminal. We can kill starlings, but we can't keep them. Rescue. Theft.

We named the bird for the Latin word meaning *song*, and Carmen joined our family—her flock—with warm and trusting enthusiasm, wanting always to be with us, on us, inquisitive about our doings, a participant in the round of our homelife. Shining, mischievous, playful, singing—flapping now from my shoulder to my wrist as I write these words.

I had a preconceived notion of what I would learn from this personal bit of starling research. Starlings are curious, intelligent, iridescent, beautiful; they have a complex social structure and are capable mimics. And so I guessed: Carmen would join me in my studio and get into so much trouble that I would marvel at Mozart's having ever gotten any work done at all. She would charm my family by mimicking our voices. She would invite me and those who met her to explore the cognitive dissonance involved in being conservationists who are enchanted by this individual of an ecologically disdained species.

And yes. Yes, she affirmed all those things with panache. Then, having dispatched my expectations, she went entirely off script, ignored my research needs, took over the story, improvised new hypotheses, composed her own results. She taught me things about the measure of my human complexity, and even more about the expanse of my ignorance. I have been thinking and writing about beyond-human kinship for more than two decades. Now a common, invasive bird perched lightly upon my shoulder and sang into my ear, *You know nothing*. Winging into the great cloud of my unknowing, this one starling has taught me ten thousand things. Here are two of them.

The Infinity of Intelligences

Because of starlings' detested status, most people are uninterested in their astonishing natural history, and even those who identify as birders have little idea that starlings are gifted mimics, able to imitate novel sounds and build a repertoire of new learned vocalizations throughout their lives. Starlings skillfully imitate other birds, cats, environmental noises, various kinds of machinery, cell phones, music, and the human voice. Rather than attempting to teach Carmen specific words, I wanted to see how her mimicry would unfold within our household, unprompted. Starlings are flirtatious, social beings, and they respond to interaction, so it was fitting that Carmen's first word was *hi,* followed swiftly by *hi Carmen, hi Honey,* and *c'mere,* the phrases we most often speak to her. Eventually, she mimicked the creak in our old wood floor and practiced the song of the Bewick's wren nesting outside her window until I couldn't tell the two birds apart. All of this was a delight—but unsurprising for a starling. Both male and female starlings sing (uncommon in female passerines) and mimic. But it took me months after she came into her full voice to figure out the most wonderous dimension to Carmen's aural echoes—that they are in truth not echoes at all.

In the dark of morning, before anyone else in the house is awake, I pad downstairs in my pink sock-monkey pajamas. As soon as I reach the bottom step, Carmen calls in a soft, whispery voice, *Hi Carmen*. The first words she hears each day. Our elder tuxedo cat Delilah follows me, ready for her breakfast—Carmen looks at her and says, *Meeooow!* in a demanding, hungry-kitty voice. I pick up the jar of coffee and, hearing the tinkling of the beans, Carmen calls *ker-klunk*, the sound of the jar lid hitting the countertop, then a gritty *whiiiir!*—not her prettiest vocalization, but an exact imitation of the coffee grinder, the sound she knows is next to come. And when I open the door of the microwave but before I press the buttons, *Beep! Beep-beep*. In rhythm and pitch perfect.

For so long I simply thought, "Wow, Carmen's mimicry is getting really good." But the moment I comprehended what was actually happening, a shiver ran from my scalp to my toes. Carmen does not just imitate the sounds of her world; she *anticipates* them, and she participates in the world by proclaiming the order of life with her voice. The more I watched, listened, and witnessed, the more it became clear that this radical aural attunement and readiness is her primary way of knowing, of learning, of communicating, and—especially, as a social bird—of sharing in the unfolding life that surrounds her.

Just as all dog owners like to think they have the smartest pup in the world, for a brief moment, I marveled that I was living with the most intelligent starling ever to rise from the stuff of creation—right here in my kitchen. It dawned eventually on my slow human brain that it is not just this brilliant little starling but *all* starlings who have such astonishing aural responsiveness to life and everything that passes within it. I threw binoculars around my neck and ran into the world. I studied wild starlings for weeks and observed this auditory alertness in the individuals everywhere around me.

Starlings are one of the most ubiquitous, most widely researched birds on Earth (in the United States, they are common lab subjects because they are unprotected, requiring no special license for collection), yet they are busy learning and expressing right beneath our noses in a manner that few recognize. The scientific literature on starlings is full of analyses regarding their vocal intelligence and the complexity of their syntax. But their anticipatory aural perception of the world is not represented in the oeuvre, which explores animal intelligence mainly by the extent to which it approximates human intelligence. Sure, we humans can hear a sound and predict what will follow. But starlings dwell in the living aural landscape as a fundamental way of being, alert in a manner beyond human capacity. And this is just one animal with one way of being, a way that I just happened to become aware of while living in uncommon intimacy with a single wild bird.

The starling's gifts are singular—as are those of all beings. Turkey vultures vocalize little, no match for a starling, but their brains house the largest olfactory sense of any bird, drawing them to freshly dead food through fragrances that rise from earth to air. My own sense of smell is trifling next to a vulture's. How must it be to live guided by fragrance and flight? What manner of intelligence forms within a life framed and molded by these things? Or the whisker-based seeing of night rodents? Or the skin-based knowing of an earthworm? Or the beyond-human echo hearing of bats? Or the rooted mycorrhizal communication of red cedars? Or the geometrical pattern recognition of bees in the flowers they see and the visual wavelengths we are blind to but that guide bee lives?

Media-driven lists of animals considered the most intelligent are most often populated by the same creatures over and over: other-than-human primates of various sorts, elephants, dolphins, border collies, crows, ravens, and parrots. The list of traits that indicate intelligence commonly include facial recognition, spatial memory, response to music, mimicry of the sounds of other animals (especially the human voice), tool use, problem solving, and grieving. These animals have eyes, most often they have fur, they live in social groups, and they do things that humans do. (The octopus, neither feathered nor furred, or even vertebrate, is trending as an outlier.) It is a positive that in recent years academic science is beginning to admit animal consciousness as a valid topic for discussion, yet both in our science and our everyday lives we continue to diminish the soaring uniqueness of other species and individuals by discussing animal intelligences only insofar as we perceive in them humanlike ways of knowing and feeling. As with the wild aural attentiveness of starlings, we who grew up with a conventional Western education constantly fail to recognize, or even imagine, the breadth of unique animal and plant intelligences that lie outside of human manners of being.

With gratitude to Carmen, I start each day with a reminder that we walk, wondering, within an infinity of living intelligences, cradled by the reciprocity of kinship in an inspirited world that simultaneously surpasses and enfolds the limits of human knowing. We walk as if in a faerie story—every being we pass, no matter how common, possessed of both message and mystery.

The Kith of Kinship

The genesis of the common name for starling—which means "little star"—is uncertain. It may have been inspired by the shape the birds' bodies form in flight, reaching in four directions—bills, wingtips, tails tapering to the point that distinguishes them from flying blackbirds. Or perhaps the celestial scattering of iridescent, pearl-white star spots that adorn their breasts in most seasons. In either case, the name *starling* is a call from the cosmos to the earth, an embodied reminder of kinship's essence. Together we are made of the fine things: soil, blood, the sustenance of earth, and ether. Starstuff.

Carmen roosts on my shoulder, quiet. Breast settled over toes, plumes soft against my neck, a slight fluff of wings lifted by tiny scapulae formed within a vertebral bauplan evolved millennia before there were any primates at all, let alone anyone in the genus *Homo* walking the earth. Can I hear her heartbeat there so close to my ear? No, but I imagine that I can. Yet I do feel a tingle on my own scapulae, as if I may sprout feathered wings of my own.

We feel this entrainment with other beings when we allow ourselves to enter into it—leaning with bare feet against the trunk of an ancient cedar, our craniosacral fluid rising and falling with the sap. The recognition of bright lightness in our own feet when the doe leaps back into her forest shelter. The alertness in the eyes of a cottontail that makes us turn to look over our own shoulders for, maybe, an even larger predator than the rabbit perceives in us. Deep kinship invites these moments of prerational interbeing with another creature, of everyday shapeshifting.

Yet this wondrous interrelatedness leaves us faltering in the face of many species' disruptive presence to ecosystemic integrity, including the starling's impact on sensitive native bird populations. In 1957, Rachel Carson wrote a paper titled "How about Citizenship Papers for the Starling." In it she praised the species' playfulness, watchability, and the fact that one of their favorite foods is cutworms—a menace to agriculturalists and gardeners. She was right, too, in noting that starlings in North America aren't going anywhere. Despite the arsenal of tactics deployed to reduce starling populations (guns, traps, explosives, and species-specific poisons actually called starlicides), starling populations continued to grow for decades after Carson's paper and stabilized at their current level about thirty years ago. It is a surprise for eco-minded people to hear a voice such as Carson's speak in favor of starlings, but in 1957 there were only twenty million starlings, a tenth of today's population. Plus, Carson, who had a love for all creatures and was fascinated by starling behavior, would have run up against the same problem we face today. In the calculus of kinship, the starling is our relation. As humans interested in acting on behalf of a wild earth with beautiful ecosystems that maintain a semblance of integrity, we face a dilemma. Starlings belong with us in kindred continuity, but what about in presence upon the landscape? How do we balance these questions in mind, body, and heart?

Here, the little-understood word *kith* that evolved alongside kinship sheds light. In modern English—even in England where the expression "kith and kin" originated—the two words are mistakenly conflated into one meaning: our relatives, those who are close to us. But the reason the archaic phrase was formed around two different words is that they *are* in fact different. Jay Griffiths points this out in her radiant book, *Kith* (changed to *A Country Called Childhood* for an American audience who is not trusted to know the word *kith* at all). The etymology of the word *kith* is murky, most likely related to the Old English *cūth*, whence the obsolete *couth*. We are familiar today with *uncouth*—a lack of knowing, an ignorance of how to act or behave. *Couth*, and eventually *kith*, by contrast, is the known, the familiar. It makes sense that the conviviality of kith came to be associated with the relatedness of kin and, as the etymologist Eric Partridge writes, "hence, by confusion, relatives."[1]

Where kin are relations of kind, kith is relationship based on knowledge of place—the close landscape, "one's square mile," as Griffiths writes, where each tree and neighbor and crow and fox and stone are known, not by map or guide but by heart.[2] Kithship, then, is intimacy with the landscape in which one dwells and is entangled, a knowing of its waymarks, its fragrance, the habits of its wildlings.

Kinship speaks to the truth of an interrelatedness that is shared no matter how deeply we as individuals perceive this connection. (We experience this with our human blood relatives; the substance of our genetic lineage remains whether or not we know our relatives well, like them, or have any sense of what they do day to day.) And although it might be more beautiful to dwell in awareness of our kinship with all of life and to act from that center, such awareness is not required for the fact of our kinship to remain as an ecological given.

Kithship is different. It is an exacting intimacy, one born of nearness, stillness, study, observation, openness. Vulnerability. Kithship is hard-won visceral intimacy—blood cut of the thorn, bright stinging of the nettle. Knowledge of the rock where the snake suns herself and the best path around it.

Kithship is particular. Among the several things that the ecologist Suzanne Simard suggests we human animals can do to assist trees in their lives and forest making is to simply go and be among them. Simard grew up in a logging family and found her early inspiration as a child in British Columbia, when she would lie "on the forest floor and stare up at the tree crowns" of the ancient Douglas firs and western red cedars.[3] It is only by dwelling over time with a *particular* forest that we can understand its uniqueness, what it needs to flourish and to thrive—and it is how, in our graced interconnection, we ourselves flourish and thrive in response. The place-based particularity of kithship explains why starlings are beloved in the United Kingdom and across Europe, where they are native. With the loss of agricultural land, their numbers are falling, and they are officially listed as a species of concern. Birders and even many academic ornithologists in the United States are stunned by this news—unable to imagine a world in which starlings are welcome. When I spoke about this subject to audiences in Austria, people were astonished to learn that the species is so despised here.

The endangered orcas in the southern waters of Scotland are my beloved kindred to be sure. I know this even though I will likely never see them. But the Salish Sea orcas who roam the home waterways we share? I know them as individuals by the scars visible on their dorsal fins. I have seen their young breach the surface of waters I paddle in my kayak. I have watched the fountain of their exhale, the echo of their breath singing all the way to shore. I have walked home after such moments in wonder, wanting never to bathe again but to live always in a skin of orca breath.

Kithship crosses dimensions of knowing that bring us to intimate specificity: book learning, alert wandering, knowledge of species close to home and recognition of individuals within these species, knowing who lives where and why, knowing who is flourishing and who is failing. Kithship enlivens and complexifies kinship, and it is essential if the fullness of kinship's wisdom is to be lived.

And the question of starling presence upon the landscape? What we should think, how we should relate, what we should *do*? Ah. I don't know. Of course, I do not know. There is no one answer, no single right response. Dwelling with kith and kin awakens, always, an unsettled complexity. With our intricate human-animal minds, we can hold many dimensions of thought at once, and such complexity is not the same as contradiction. We are asked to walk lightly and intelligently within an essential ambiguity.

In kinship, Carmen and her own kindred starlings with their ravishing intelligence are my relations—sister, mother, beloved. In kithship, I pause. I observe the flicker who was evicted from her nest on the corner by a starling, recognizable by her habit of roosting on a particular cherry branch, near the tree's trunk. I wander the woodland edges near my home where starlings do not nest and witness the uptick in native avian biodiversity there. I behold the starlings who swirl from their exquisite murmuration out of the sky and into our backyard fir. I watch Carmen when they begin to whistle; she falls silent, tilting her head. I wish starlings were not present upon this landscape. I know that killing them will not help and is unjustified. I know, too, that I cannot accept their presence with a full heart. We stand in a glorious, tangled dissonance filled with love, intimacy, and confusion. I cover the holes in my home where starlings might nest. I plant trees. And when I see a starling? I stand in awe of her loveliness and whisper, "Hello, shining one."

NOTES

1. E. Partridge, *Origins: A Short Etymological Dictionary of Modern English* (New York: Greenwich House, 1983), 73.
2. J. Griffiths, *A Country Called Childhood: Children and the Exuberant World* (Berkeley: Counterpoint, 2014), 6.
3. S. Simard, "How Trees Talk to Each Other," video filmed June 2016 at TEDSummit, Banff, AB, 0:49, https://www.ted.com/talks/suzanne_simard_how_trees_talk_to_each_other?language=en.

WHEN ONE KNOWN TO YOU DIES, THE REARRANGING OF SPACE AND TIME BEGINS

Elizabeth Bradfield

*For Ladders, 2019 (*Balaenoptera physalus*)*

A rib (I know whose) in
the harbor under
waves. How

heavy would it be, hefted? Low
tide will bare it,
will allow

pickers to take it, make
it décor—whale bone
with tulips, leaching

minerals, oil, the perennials
stronger for it. Up
the beach

the rest of him. Un-
scattered and held still
by sinew, flesh.

Spine and ribs but no longer
the jaw, which when he first
washed up ashore

and was flayed by flensers and sun,
proved to be broken.
We knew him,

this fin whale, Ladders. I can't remember
the year, the moment of
my first

sighting or resighting of his stuttered
prop scar, long healed, an easy
marker. Who

was I then? Young and newly
arrived, sorrows vast, and
losses, it must

be said, negligible. Negligible.
What does his rib curve
now? That

space filled by water. That
emptiness. And knowledge
of what it once held.

SKINFOLK, KINFOLK, AND THE KINSHIP OF ONENESS

Orrin Williams

People usually consider walking on water or in thin air a miracle. But I think the real miracle is not to walk either on water or in thin air, but to walk on earth. Every day we are engaged in a miracle which we don't even recognize: a blue sky, white clouds, green leaves, the black, curious eyes of a child—our own two eyes. All is a miracle.

—THICH NHAT HANH, *THE MIRACLE OF MINDFULNESS*, 1999

The concept of race is a recent concept and social construct in human history. Many sociologists and biologists do not see race as real. Yet it is real in terms of the damage done to people, particularly those who are already marginalized. It also damages those who perpetrate social hierarchies based on race and materially benefit from this caste system. The concept of race has propagated a social system rooted in a color code that I have begun to refer to as a system of "skinfolk." By virtue of having very recent ancestors of African descent, according to this social construct, I belong to skinfolks often described as the "Blacks."

"Blacks" as a skinfolk category did not always exist, and it is distressing that contemporary notions of race and skinfolk are now retroactively used to describe the panoply of humanity throughout history. One example is the oft-repeated statement that Blacks sold Blacks into slavery and were therefore complicit in the perpetration of the Atlantic slave trade that brought millions of "Black" people to the Western Hemisphere or the Americas. During the period of the Atlantic slave trade, however, there were no "Black" people. Instead, there were ethnic groups that did not recognize divisions based on skin color.

Nor did color similarities—what might be called "skinship"—translate into kinship. The phenomenon of ethnic conflict is, of course, not unique to "Black" people. Ethnic conflict is a profoundly destructive part of human history and interactions. Lest we think that conflicts between skinfolks are confined to Black skinfolks, one need only consider the historical religious strife in Northern Ireland, the tension between the Walloons and French in Belgium, the competition for supremacy in Asia between China and Japan, or the conflict between Muslim and Hindu skinfolks in India. Contemporary conflicts also continue to occur between people whose skinship is presumed to be the same; for example, conflicts between ethnic communities rage across Africa, persist between Haitians and Dominicans, and manifest on the streets in urban areas large and small across the United States.

The profound tragedy that my skinfolk have experienced since the inception of the slave trade and subsequent periods of racialized oppression is well known. The process of forced migration during the enslavement period resulted in Africans from different ethnic groups being separated and thrown together as a way to reduce the likelihood of revolt. This was also a critical part of an attempt to destroy diverse African cultural systems and beliefs.

Yet as it is sometimes said: in crisis appears opportunity. As cruel as the attempted erasure of African cultures was, this intermingling of African ways of life also opened opportunities rooted in skinship, including social and cultural concepts such as pan-Africanism, which espouses positive Black cultural ideas and seeks to overthrow negative patterns of thought associated with blackness. Cultural consciousness is an antidote to the destructive aspects of enslavement, oppression, and colonialism—a pathway to healing from the intergenerational traumas experienced by those identified as occupying enslaved, oppressed, or colonized classes.

As a platform for social cohesion, skinship could be critically broadened by focusing on kinship. What is apparent is that a (r)evolution is in order, and it must be waged via a campaign that is rooted in the recognition that all people are sacred beings and thus kin. Pan-African and Black consciousness movements contain nascent ideals that open fresh pathways for considering what kinship means. We can no longer afford to be in perpetual conflict with one another. It does not matter which sector is evaluated, from the economy to spirituality. The divisions that perpetrate war and hate must be ended; kinship must prevail if healing cultural and ecological trauma is a priority.

Just as a lack of kinship recognition among peoples leads to inevitable conflict and attempts at ethnic domination, the current dominant economic "system" reveals a lack of kinship recognition between human and nonhuman worlds. This system is extractive at every level, perpetuating industries that mine everything—from the fish in the ocean to rare-earth minerals to fossil fuels. Extractive social interactions in Western social systems mirror patterns of extractive practices such as mining. Moreover, extractive social interactions and extractive economies often reproduce the same negative biological, social, cultural, and psychological impacts.

Marginalized and oppressed people and communities are particularly susceptible to extractive social mining, which manifests in societal pollutants such as the "drug war," extractive education systems, standardized testing, and the entertainment complex.[1] If marginalized and oppressed people "perform" well, on the basis of an extractive separation process, they are selected for professional, technical, and scientific positions. More often, they are placed in social strata where they are subject to the prison-industrial complex, retail and warehouse jobs, and gig and underground economies. For whichever category one is selected, the outcome of marginalization remains. This happens regardless of titles, awards, and achievements. (In this respect, one might note reactions to the Obama presidency; or the assassination of Nobel Prize winner Dr. Martin Luther King; or the barely recognized contributions of Dr. Mark Dean, a person of African descent who invented the personal computer platform that we recognize today.)

Similarly, many people captured and transported during the enslavement period were selected because of their agricultural knowledge of crops such as rice and indigo. One result was the huge agricultural region known as the "rice bowl" that thrived in the southeastern United States, which, at least in my mind, is an early example of intellectual property theft. This led to the enrichment of some at the expense of marginalized and enslaved people whose well-being was treated as less important, regardless of their knowledge. Such enslavement, marginalization, and extraction among humans finds ongoing echoes in how the rest of our ecosphere is treated.

My cumulative study, exposure to, and practices of Buddhism, Taoism, Rastafarianism, Traditional African Spirituality, and other global Indigenous shamanic systems have formed my commitment to kinship. This was fueled by episodes of realization that go back as far as I can remember.

Those episodes of realization began when I was only six years old. On a family trip to visit my great-grandmother in Florida, I was with my parents and grandparents and we wound up at the site of a carnival. I remember wanting to go to the carnival, but I was told no. I whined, as children often do, until my grandfather finally said, "Let him go." I got in line, although no one else from my family followed. When at last I reached the entrance, a white man—who I can still visualize almost sixty-five years later—told me, "Niggers can't come in here." As I walked back to my family, I vividly recall my reaction, which was a resolution: I would begin a journey to unravel why I was treated that way and what could be done to ensure that no one would ever be treated that way again. I also remember my grandfather's gaze. He had imparted a gift of wisdom—one that keeps giving to this day. It is manifest in gifts such as this piece I am now writing and in the admittedly imperfect ways I try to live a life of peace, love, and interconnection.

In contrast, on that same trip, I remember running up and down beaches on the Gulf of Mexico chasing crabs and accompanying my great aunt to the pier every morning to purchase fish and seafood as the fisherfolk brought in their daily catch. I remember being barefoot virtually all the time, except to go to church or another event where (I guess) shoes were required. I tuned into the feel of the warm earth encircling my feet. I remember the morning rain and the afternoon heat.

I also remember pockets in Chicago that held on to a bit of the wild. Near my grandparents' building, I spent quiet, secret times with the dragonflies, tadpoles, and reeds.

When I served in the US Air Force in my twenties, I was sent to Thailand. Before that, my only experience with "Orientals" was in Chicago's local Chinatown or at the Chinese laundry near my home. I thought Asia and China were equivalent. My brain exploded in Thailand when I encountered various local people, some of whom were darker than me with hair nappier than mine. This diversity was expressed in the food, as well, which was not "Chinese" but a reflection of the mélange of humanity I encountered. The ecology of the country also awakened my connection with the natural world, which was solidified later by a general ecology course that helped me recognize clearly the web of life and how I was part of it.

In more recent years, episodes of realization about my true interconnected self occur particularly during the early morning hours of three to about five, when the veil between the cosmic spirit realm and the earthly realm is the thinnest. Sometimes I will tune in to my breath and recognize that it is a miracle, that my life and that of all beings is a miracle, and the word *miracle* becomes a mantra recited silently with each exhalation.

When I am looking out of the window of a building or when I am in the car, I sometimes begin a spontaneous mantra of "interbeing," repeating that word silently with every being I see, be it a human, bird, dog, or rabbit.

Similarly, spontaneous episodes of realization occur when I sit in my garden, or at a park, or a farm, watching the gathering of birds and pollinators while connecting with each plant spirit. Then the mantra "gratitude" emerges from somewhere deep inside, forming a chorus with the singing birds and other ambient sounds.

Every instance brings the potential realization of kinship and interdependent origination. As the Buddhist and clinical psychologist Jack Kornfield reminds us in the title of his book *After the Ecstasy, the Laundry,* whether we must go to work, wash the dishes, or clean up, when considered deeply and mindfully, these are opportunities for fulfillment.

Observing the world from a broader vantage point, or what I sometimes call my "cosmic perch," I see my beautiful brown skin not as a symbol of caste, enslavement, inferiority, or oppression but as a remarkable expression of resilience, adaptation, and expression of DNA. That DNA clicked on to protect my ancestors from the rays sent forth by Grandfather Sun as he provided valuable nutrients such as vitamin D to our naked tropical bodies. My skin is thus a tribute to a profound cosmic and sacred connection to the oneness of all things.

Thus, I proclaim that we are cosmic stuff, sourced from the same stuff that forms all actions and manifestations of the Cosmic Mother-Father. The cosmic flow is in its totality incomprehensible, yet, sacred and profound and in that knowledge, I stand in awe as our hominid ancestors must have done. As such, I find the source and wisdom for my kinship with all beings, be they microscopic, plant, animal, soil, or rock.

As I look out at the world, from a cosmic perspective I see remarkable diversity that has adapted to life on this planet. The humanscape is rich because as kin we have adapted to live in a multitude of lifescapes—from the Arctic Circle, to the mountains, to the savanna and the forests—drawing sustenance from the water sources and soils where we evolved and adapted to the ecosystems and bioregions where our ancestors settled.

At present, most of us no longer live in isolated patches of community. We are at some level aware of one another and our remarkable diversity. I can say with certainty when looking deeply from my cosmic perch that I see kinship with all the humanscape. Sometimes it is difficult to reach that level of wisdom and discernment when faced with the various social constructs and interactions we currently experience. Yet at a deep spiritual level, I can see who I am, and myself in others, as kin.

When I connect to oneness—or "interbeing," as the great Dharma teacher Thich Nhat Hanh calls it—I can shake off any notion of insignificance. Instead of feeling as though I were a single grain of sand on a vast beach, I am filled with a sense that I am part of a grand system, and my connection to it means that I am an embodiment of that system. The same conscious connection to the cosmos elicits feelings of oneness in me. At the level of being a sentient being on Mother Earth, I therefore see myself as being in kinship with *all* sentient and nonsentient beings on this grand planet.

The task at hand, as I see it, is how we might craft a consciousness of kinship and oneness that allows us to see each human, plant, animal, river, lake, ocean, and microbe as a sacred whole. From this consciousness, new and evolving systems of stewardship, rights, economies, and practices that are rooted in biomimicry can arise.

The work of multiple organizations that I have the honor to work with reflect the movement toward shifting paradigms attuned to a (r)evolution that is both human and ecologically oriented. They also offer opportunities to reject paradigms that separate humans from the "other," moving humans directly into integration with global ecosystems and kinship with all beings.

For example, the Sweet Water Foundation, with its dedication to sustainable farming practices and providing a portion of the farming yields to community members who have little income, is a project devoted to land stewardship and a just economy.

Similarly, Growing Home is the only certified organic urban farm in Chicago. Under its leadership, it is devoted to increasing the amount of produce available to a community lacking access to healthy food and experiencing food insecurity. This initiative is the epitome of systemic and environmental changes, in which the food is grown at a price that reflects the economic reality of the Greater Englewood community in Chicago.

Likewise, our work in partnership with the Stein Learning Garden at St. Sabina and the Chicago Grows Food initiative are committed to expanding the number of people and households that grow some of their food. One of the salient points in our partnership is using gardening to reconnect people not only to their food but also to our relationship to Mother Earth and Father Sky.

These local initiatives align with many initiatives around this amazing planet to preserve our fisheries, protect our soils through regenerative farming, and clean our air and waters, thus actualizing the common admonition to "think globally, act locally."

May the cosmic dance—the choreography of planetary and cosmic unfolding—continue. May that be the dance that determines our future rather than the machinations of a handful of imperfect beings who are determined to act in disharmony with the divine, cosmic order that is the source of our existence.

When I'm sitting on my garden perch instead of my cosmic perch, I am well aware that I live in a world shaped by various skinships. I struggle within and against such constructs that unthinkingly or deliberately aim to keep myself and others "in our place."

Those kinds of racialized skinships are oblivious to the more fundamental kinships we share as fellow earthlings. We must resist their extractivist logics and oppressive forces. On the South Side of Chicago and beyond, I work toward practicing these kinships. And sometimes, looking out from my garden perch, where I can see my kin all around me, a familiar mantra repeats in my head: *gratitude,* bumblebee; *gratitude,* flower; *gratitude,* birdsong; *gratitude,* soil; *gratitude,* human kinfolk; *gratitude,* interbeing. I sometimes think of the little boy at that carnival entrance, stung by hateful words of division. I respond: "I am not simply a Black boy; I am a being 4.5 billion years old."

NOTES

1. Regarding the so-called drug war, see the Equal Justice Initiative website (https://eji.org/news/nixon-war-on-drugs-designed-to-criminalize-black-people/) and James Forman Jr., *Locking Up Our Own: Crime and Punishment in Black America* (New York: Farrar, Straus & Giroux, 2017). On the extractive educational system, see Jonathan Kozol, *Savage Inequalities: Children in America's Schools* (New York: Broadway Paperbacks, 2012); Jonathan Kozol, *The Shame of the Nation: The Restoration of Apartheid Schooling in America* (New York: Broadway Paperbacks, 2005); Paulo Freire, *Pedagogy of the Oppressed,* 4th ed. (London: Bloomsbury, 2018); Carter G. Woodson, *Mis-education of the Negro* (New York: Tribeca Books, 2011). On standardized testing, see John Rosales, "The Racist Beginnings of Standardized Testing," http://www.nea.org/home/73288.htm; Richard V. Reeves and Dimitrios Halikias, "Race Gaps in SAT Scores Highlight Inequality and Hinder Upward Mobility," Brookings Institution, February 1, 2017, https://www.brookings.edu/research/race-gaps-in-sat-scores-highlight-inequality-and-hinder-upward-mobility/; Christine Brigid Malsbary, "Standardized Tests are a Form of Racial Profiling," *Common Dreams,* October 26, 2015, https://www.commondreams.org/views/2015/10/26/standardized-tests-are-form-racial-profiling.

BADWATER

Daegan Miller

—A quick prickle of skin.

My entire High Plains life is piled neatly in the dirt, there on the ground beside my shirt; I'm in shorts and boots, only. My eyeglasses, thick, lie on a flat rock nearby, so that I can pour pan after pan of cold well water over my dusty head, water that licks cleanly down my shoulder blades and into the small of my back before raining onto the cracked earth. I've been out of potable water for hours, I am parched by the heat of a scorching July day, and I can feel each drop as it hits my thirsty, elated skin. Then a prickle, colder than the breeze: something comes through the sagebrush. Something coming, but the world is smeared in my nearsightedness.

It began like this: I lied when the ranger asked if I had packed in enough water, a gallon per day, for my three-day trip—all the water out in the Badlands was bitter, he said, too alkaline to drink, and the Little Missouri was so silty that it would clog a filter. I also chuckled when he warned me about the buffalo. "If you see one, don't run," he said. "They're nearsighted, it's their mating season, and the males get a little crazy this time of year. If you know what I mean." I was a little crazy, too: twenty-four, as wild and free as I've ever been, fresh from a monthlong solo hike of Vermont, I was attending a National Endowment for the Humanities–sponsored, six-week summer seminar at North Dakota State University in Fargo on the history and culture of the Great Plains, although at some point I decided that I had had enough of the seminar room—I had lived nearly my whole life in the Northeast, and I needed to get out, on foot, away from the chaos of concrete. To see the land at night. To measure its length against my body, hear the insects and the sound of an easy wind, smell the prickly smell of silver sage at dawn, taste noon's dust, and learn, with my skin, the way the sun burns its way across that immense sky. I wanted an adventure, and so I packed my bag for the Badlands protected by Theodore Roosevelt National Park, in far-western North Dakota: food, stove, and three quarts of water went into my pack, along with my copy of Wallace Stegner's *Wolf Willow,* an account of growing up on, and never quite fitting into, the plains. I packed my bag, told a little lie, and was off.

Suddenly cold, I grab for my glasses—slap them to my face, watch as a thirsty buffalo emerges into the clearing at whose center lies the well from which I drink. Behind the leader, the sage rustles and tosses with the bulk of a few dozen dark-brown humps. I glance at my gear on the ground and then back to be fixed in the gaze of the buffalo's eye.

Theodore Roosevelt came to the Badlands in 1883 to hunt buffalo; he had been an asthmatic, sickly child, and so he was a devotee of the rugged life. Shooting and fishing and fighting—these, he thought, had beaten the softness from his body, had hardened him into a man. When he arrived, he was stunned by the austere beauty of the landscape—"the ground is rent and broken into the most fantastic shapes"—as well as the craggy beauty of the cowboys whose vitality he coveted. "Sinewy, hardy, self-reliant," he wrote of the white men he encountered in North Dakota, "their life forces them to be both daring and adventurous, and the passing over their heads of a few years leaves printed on their faces certain lines which tell of dangers quietly fronted and hardships uncomplainingly endured." Roosevelt was there to front danger, but he went mostly luckless and bored, until, only a few days from his departure, he spotted his prize: "His glossy fall coat was in fine trim, and shone in the rays of the sun; while his pride of bearing showed him to be in the lusty vigor of his prime," Roosevelt wrote of the buffalo he finally encountered.[1] And then he shot it. It was hard work cutting off its head, but he wanted a trophy to show his friends back east. Today, you can walk into his estate, Sagamore, on Long Island, and see it for yourself: the buffalo's head is still there, hanging on a wall. Roosevelt was twenty-four and looking for adventure.

A deep, living, liquid brown; a wild brown, a free brown, a color all its own; an eye that flickers and sparks, that takes me in as I stand; and that is how we remain, eye to eye, in the sage- and dust-scented air of a Dakota summer day.

Imagine a square at whose center is Theodore Roosevelt National Park, which was created, in 1947, as a memorial to the twenty-sixth president, who, after his hunting trip, bought himself *two* different ranches, parts of which are still preserved, in the vicinity of today's park.

"During the past three centuries," Roosevelt wrote in 1889, "the spread of the English-speaking peoples over the world's waste spaces has been not only the most striking feature in the world's history, but also the event of all others most far-reaching in its effects and importance."[2]

Imagine that at each of the square's vertices is today a reservation. Fort Peck (established in 1871) in the northwest. The Crow (1868) and Northern Cheyenne (1884) reservations below it. In the northeast, Fort Berthold (1870). And in the southeast, Standing Rock (1889) and Cheyenne River (1889).

It was this violent taking of "waste space," Roosevelt was sure, that made American men "the mightiest among the peoples of the world," and it explains why he spent so much time, in the mid-1880s, playing the life of a frontiersman out on his ranches.[3]

Seven different pens to build a nation: six to keep people in, one to keep them out.

By 1901, when Roosevelt assumed the presidency, white American men had taken the country, had killed and cleared the American Indians from most of it, had decimated the buffalo—there was nowhere left to go and little left to shoot. So Roosevelt, the conservation president, would come to preserve more than 234 million acres of land, in part, to preserve a place to roam, prevention against "a certain softness of fibre" that came with too much civilization, and so ensure the supremacy of white American manhood, forever.[4]

"Holy fucking shit," I gasp, and start backing out of the clearing, backward into the sage. I will marvel, later, at how its branches scrawled my bare trunk and limbs. Slowly, I back away, as the ranger had advised. The buffalo, joined by his herd, follows; they all do, ambling faster than my creep, so I turn and stumble into a brisk walk, which quickly becomes a trot, then an agitated gallop, and finally a mad, zigzagging sprint through the sage. A paved park road is about a hundred yards ahead and is jammed now with idling car-bound tourists who had also caught sight of the herd; it's toward them, toward my own air-conditioned kind that I bend my course, tearing through the brush, waving my arms and hollering to be let in, hearing instead, once I draw close, the thump of locks slamming down into their catches.

I once before had stood looking into a buffalo's eye, an experience that came mothballed in four generations of family myth. The backstory, as I remember it now, thirty years after my grandmother told it to me, is that my great-great-grandfather, a German immigrant with a sharp eye for opportunity—he supposedly owned the first roller coaster in the United States and was asked to tour with Buffalo Bill Cody but turned him down because he thought Cody a crook—knew Roosevelt in passing, had talked hunting with him, had shown Roosevelt some of his stuffed trophies, and had killed the largest buffalo ever recorded. The proof was hanging on the wall of the family barn in the Catskill Mountains of downstate New York. I remember one day sliding the barn door back on its ancient, pitted iron track and—there it was. I remember being awed and confused by something so big, so foreign, severed from its context in the dim light and moldering smell of the barn. I remember looking into his eye and seeing nothing at all except for my own prickling sadness, reflected in brown glass.

Off to my left I see, out of the corner of my eye, a tree, not much more than a bush, that I'll remember later as some sort of scrubby, prickly oak, not much more than a sapling, not much taller than I am. But it's all there is, and I make for it, for its lowest, stoutest branch, of two fingers' thickness, an easy reach, and in a moment,

I've swung up onto it, sitting as close to the trunk as I can, hugging it, squatting on the branch as it bows groundward but holds. Holds me. Then they're underneath.

Who am I, where do I belong, with, and to, whom?

Am I the historical narrative I was born into—and if I am, which events, people, places, and beings are my kin?

Or am I the relationships I choose to make and break, the relationships that will make and break me, those acknowledged, those unknown?

What do I owe the past—to my great-great-grandfather and to Teddy Roosevelt? What do I owe to the American Indians whose lands and lives were taken, whose lands and lives yet persist? What do I owe the still-living sage and the scrubby oak?

What do I owe the buffalo head nailed to the wall in my family's barn?

The entire herd drifts beneath and around the oak tree as I perch in its branches, like a strange, flightless bird. They're so close that I feel their bodily heat against my belly. I taste the muskiness of the dust thrown up from their skin, can see a lone fly strut across a hump; I can feel them low, can feel their hooves plod up through the tree's trunk and ripple through the leather of my boots. Their smell pricks my nose—the rich smell of fecundity: sweat, earth, and life, a wild smell that envelops the tree. If I wanted, I could dip the toe of my boot just a few degrees and stroke the back of the tallest. But I don't. It would feel like a violation, and besides, everything is perfect right now. Everything exactly as it needs to be, as I hover a whisper above the buffalo who disappear into the sage.

I sit in the tree for a very long time, long after the herd has passed, crossed the road, and wandered up over the next rise and out of sight, after the cars have released their emergency brakes and dispersed to wherever they were going before they stopped. It's only once the last brown sniff of buffalo has settled into the dust that I climb down and walk, slowly, back to my gear. Not a thing of mine has been touched. Not my camera, which is safe, not my food, which is whole. My uncapped water bottles remain full. Not a thing has been touched except for my book, Stegner's *Wolf Willow,* which bears a muddy hoof print on its cover.

It's been almost twenty years since my visit to Theodore Roosevelt State Park, but I can easily recall the feeling of staring deep into the eye of another being vastly different than me, of seeing in that eye something wild, unknown, almost familiar, and feeling its deep vitality; of feeling small, insignificant, vulnerable, and so dependent. I have a few pressed flowers in my notebook from that trip—an aster, a prairie rose, a sprig of sage, which still smells, if I hold it just right, slightly of life. If I close my eyes, I can feel the unconditional support of that spindly tree that shouldn't have held me but did.

On the last night of my trip, water bottles full, I camp on a bluff overlooking the Little Missouri River, opposite an RV park, and feel the purples of the high prairie darken to black as I strip off all of my clothes, skin prickling, arch my back, and howl, full of thanks, long and loud and free, out into the night. From the campground comes shouted swearing, but then a chorus of dogs barks in sympathy, and, way back in the sage-silver hills, the yip of coyote song.

NOTES

1. Theodore Roosevelt, *Hunting Trips of a Ranchman: An Account of the Big Game of the United States and Its Chase with Horse, Hound and Rifle*, 2 vols. (New York: G. P. Putnam's Sons, 1885), 1:3, 8; 2:105–6.
2. Theodore Roosevelt, *The Winning of the West*, vol. 1, *From the Alleghanies to the Mississippi*, 1769–1776 (New York: G. P. Putnam's Sons, 1889), 1.
3. Theodore Roosevelt, *The Strenuous Life: Essays and Addresses* (New York: Century Co., 1901), 249.
4. Theodore Roosevelt, *American Ideals and Other Essays Social and Political* (New York: G. P. Putnam's Sons, 1897), 336.

CORNFLOWERS

Brenda Cárdenas

She says my hair smells
like corn tortillas.
I raise an eyebrow.
After all those honeysuckle
and papaya shampoos,
I can't believe my scalp hasn't soaked up
the scent of blossom
or the perfume of rainfall.
No, she's my mother,
and she insists
that even as a little girl,
my whole bedroom breathed
corn tortillas.

Pressing nose to pillowcase,
I search for masa,
reach back before molcajete and plow
to a dusky meadow,
its bed of soil flecked with teosinte,
ancestor grasses.

Up through the dark follicles of my skull
covered in sun-cracked husks,
push the black-brown silk strands, cocooning thirsty kernels.
Maíz sprouts into fields of thought bearing hybrid rows of words
that fall like teeth
from the mouths of the dead.

SKINCENTRIC ECOLOGY

Andreas Weber

After rain I run my hand through juniper or birches for the joy of the wet drops trickling over the palm.

—NAN SHEPHERD, *THE LIVING MOUNTAIN*

The Minerals' Skin

Every time I looked up from my writing, I saw the lichens. They covered the opposite roof. It was a low roof, not steep at all, wedged in between walls of gray stones, covered by reddish clay tiles.

At first, when I found my writing place here, at the small window looking over the other building, I had not noticed the lichens, mistaking them for weathered patches. These beings—composed of algae and fungi in a single organism—formed rounded spots and spherical halos on the mineral surface. There were black spots, and dots made of soft gray, and circles that in their reddish hue seemed like transformations of the clay itself. I let my eyes wander over the roof made of mineral. Before my gaze, the lichens' patches transmuted the surface. The mineral bloomed, and its blossoms were slowly spreading, touching one another, growing into each other, meandering around the spaces in between, bleeding into one.

Where the lichens dwelt, the texture of the surface softened and seemed almost creamy. My eyes softened, too. My gaze grazed the lichens, and I felt as though I could ingest the stone, which had become palpable, touchable, edible. My vision worked in two directions: by watching the lichens softening the stone, I was touched by them and mollified by their touch. They gave back my gaze, and receiving theirs made me soft. The lichens were a dreaming of the rock. And I was a dreaming of the lichens.

I had a lot of time to watch the roof. I was alone in a silent house, spending some weeks between rows of olive trees in the hills of Tuscany, high above Siena, to take care of the cats of a friend's friend. After my arrival, the succession of moments gradually slowed down and left me with two companions to reflect upon: stone and being. It was January, and in the night the temperature outside dropped below freezing. On some mornings, the high ridges to the southeast of Siena transparently shone through the mist.

The discovery of the lichens' presence instilled not only pleasure in me but also a sense of urgency. It was a sort of yearning, as though I should not waste a minute and pay due attention to what they gave to me. They watched me with the gaze of living stone. They sat there, on that roof, as part of that roof, as stone that, if you wait long enough, softens and becomes palpable as a living skin.

Admiration and Grief

Whenever I raised my eyes to the lichens and their spherical patterns on the tiles, I experienced a profound bafflement. I had strong feelings but no words to express them. A clear, sharp beauty slipped through my fingers. The lichens were there, plainly there, just there, present, unmoved, soft and dimly shining, like stone undone—and, at the same time, distant, closed into themselves—making me feel that I could not reach them. I could feel the pull of their presence, but it left a void.

I thought of Thomas Mann's characterization of love as a mixture of "admiration and grief," a sentiment I had never liked, since it seemed to be a narcissistic misunderstanding of connection. But I felt something similar here—and feeling it made me more uneasy. I was held in deep attraction—and in endless distance. After all, those were just tiles with epiphytes on them. Why did I experience their presence as meaningful? I gazed upon them and felt watched. How?

My uneasiness was not just about my personal state of mind. It had to do with something more general: with how most humans relate with other beings and how we share our world with them. I felt bad because of the rule that we humans ultimately are strangers to other beings—to lichen-beings, tile-beings, algae, minerals, water, the stones of the blue Sienese hills. Aren't these just things? When they suddenly speak, we are startled. We don't know how to respond. We don't know how to welcome back. We are unhappily in love.

Sitting at my window overlooking the lichen blooms on the roof, I was too overwhelmed by what I felt on my skin to give in to this resignation. Although I was only watching them, the velvet spheres out on the roof made my body tingle. They made me joyful, nervous, and restless. It was a view of other beings' skin. The voice inside that whispered to love back did not originate in my head; it was my skin murmuring. It was the soft and touchable aspect in my own flesh that answered. My flesh could not remain indifferent to being touched.

What let me open up to the lichens was that which was lichen inside of myself, slowly softening the surface of a stone and making it blossom with a velvet epidermis. What responded within me was me-the-lichen, me-the-algae, me-the-mycelia, me-the-rock. Its whispering came not as a pertinent voice but as a gentle touch from the inside, a sweetness that appeared and disappeared and came back again, in waves of making and unmaking, like a tender breath.

I realized that the slight taint in the beauty I experienced came from not giving in to my own desire.

Breathing Together

After lunch, I used to walk up the slope behind the house. My companion, a black poodle, raced ahead, happy to move—although she seemed equally happy indoors, lying on the window ledge looking out, keeping company with lichens and hills. The sun was high. We strode past oak trees stretching their barren branches into a transparent sky. We trod over crisp brown leaves, along the withered manes of last summer's grasses. High in a tree, hunters had set up a shooting platform constructed out of crinkled boards and camouflage tarp that silently awaited the arrival of the songbirds in spring.

Along the path, granite boulders pierced the earth, softly rounded mounds of gray and white, orange and black. They consisted of stone and flesh, as did the roof tiles. They were covered with dense crusts of lichens. The sun was warm. It had chased away the hoarfrost, and now caressed the stone with careful rays. At the ridge of the hill, where we paused before turning back, a massive boulder rose up from the earth like a colorful cupule, overgrown with vegetation. On top of the boulder, a whitish circle rippled outward in waves, like a radiating sun.

While I sat on that massive stone, careful not to scar the lichens, I found it difficult to tell where the lichens started and where the stone ended. Both had become one being. And indeed, as the lichens grow on the mineral surface, they feed from it. They extract minerals and incorporate them into their bodies. A rock that is colonized by lichens weathers a thousand times faster than it does if it is not embraced by life. Lichens eat rock—just as they eat sunlight. They transform rock into flesh. Their flesh is mineral.

I sat on top of the hill and watched the minutes sink slowly into the blue of the distant valleys. I caressed the coarse rocky skin with my fingers, allowing our skins to merge. I lingered in the presence of the lichens, touched by the lichens, as skin among skin, as breath from the rock's breath.

Metabolism is a way through which one being becomes incorporated into another, not metaphorically. Metabolism is a way stone becomes me. What in my heart felt like an exchange with plant beings and fungus beings and rock truly *was* this exchange:

plants transform rock, and, by this pathway, my body—as I subsist on plants, like all life does—transforms from rock into flesh. The same sort of transformation happens as I breathe. I breathe in the exhalations of plants, and they breathe in my body, whose building blocks of carbon are continuously broken down and shed through my lungs as carbon dioxide.

A similar transformation happens when I eat: I convert the bodies of other beings into my own. It happens when a root digs into the soil, dissolving its grains and taking up its elements. All those are the in- and out-breathing movements of how the stuff of this world is transformed through beings who meet, touch, intermingle their skins, become one, and separate again to become others. It is all breath. It is all touch. Every incorporation is a meeting of two sensitive surfaces, an exchange of skin through skin.

In every moment, life is the birth of one being into another. I am given to myself through others, and I can go on breathing only by allowing myself to pass away into others. The lichens on that roof were a direct part of this exchange. Some of the carbon dioxide I exhaled yesterday found its way into their bodies. I looked at my own flesh and blood. We were a physical continuity. We were family. Skin is kin.

Persons of Matter

When we experience beauty, something in us knows this. Our sensible skin knows. Our breathing chest knows. Our eyes, taking in light, and radiating light outward with every gaze, know. We know that we are part and parcel of this grand exchange. We know that we are family.

I have not revisited the silent stone house looking out over the Sienese hills. But the experience has remained with me. So, still today, lichens exert their magic everywhere I look. In the forest close to my place in Berlin, they cover the trunks of the winterly barren oaks with hues of whitish green. The lichens grow on a portion of the trunk's circumference where they are exposed to a certain amount of rain and sun. On other parts of the trunk, green algae cover the bark with sulfuric yellow. The lichens have needs, and they act according to them. Often I stop at a tree and let my hands glide over the soft coarseness. The lichens are cool, and slightly moist, and they always have a tender grain, like exquisite velvet. I stay and breathe, and at some point I start to see the lichens as the selves that they are, with needs and preferences. I don't always achieve this, but when I do, then the world suddenly shifts. Every physical detail, every loop and bend of their thalli becomes a gesture of their ways of being.

We are all family because we all share the feeling of being alive. We all share ways of realizing this feeling. And we all share the atoms and molecules that embody this feeling. We breathe one another. And we perceive others striving for the same goals that we strive for: continued existence, connection with others, exchange of flesh through flesh. In the other beings' matter we can see ourselves before us, and at the same time we are this being we see there.

Our ways of being alive come about through bodies that are mutually breathing one another. At the same time, each individual's way of living according to his or her feelings is unique. And each species' tradition of fulfilling those needs is equally unique. While stroking the lichens, this insight comes to my skin literally as firsthand knowledge. Their uniqueness compels me—the sheer fact of this soft, coarse texture in its pale white, here and now. The uniqueness of a self.

Biology has shown that each being is fundamentally "autopoietic": living beings create themselves. Every breath is an act of mixing but also an act of autoconstruction. Organisms are those parts of the living flesh that show an insistence on remaining an active center, an agent, someone to whom its own being is of concern. From this biological perspective, a cell is a subject with needs. A cell is a self. A self is a person.

This is not limited to biological organisms. Organisms express a desire to be in connection, but everything takes part in desire's yearning to become through mutual transformation. Stones do. Their openness to new encounters manifests in the slow withering of their crusts. Everything temporal partakes in realizing desire. Everything that happens pushes it further. The arrow of time is the arrow of desire. Time is there because things happen, because atoms meet, because stones breathe one another. Matter is social. Time arises because this cosmos cannot sit still. It needs to share and connect.

If we need to share, then it becomes crucial to what degree our sharing allows us to flourish. If we—granite beings, lichen beings, dog beings, and human beings—need to share with others, then the transmutations of flesh into other flesh are not just silent mechanical processes but always are colored by yearning. If all of us beings need to share, this world is not a neutral place but filled to the brim with feeling. All skin we encounter is sensible skin, like our own, which through its sensibility transmits the urgency of the other's desire to change form with ourselves.

Our skin knows. Our skin even knows when it does not touch other skin directly, but when we graze the surface of another being with our eyes. Our skin knows, as it is led by the probing fingers of the lichens slowly converting the stone's longing into sentient flesh. We are matter, and we feel through it. Living through a sensitive skin is how matter feels itself.

Aliveness means to partake in the desire to be, and in the desire to connect. It means to let our skin be touched, to suffuse it with otherness, and feel through it. Membership in the desire to share makes a person. Aliveness is personal. It is addressing us personally through our skins, through which we feel the other. We exist as threads of an endlessly extended mycelium in which everything is of our own flesh and blood. At the same time, all the bliss and all the suffering are experienced by selves, by *persons of matter*, who yearn to become fuller through mutual transformation.

Beauty Is Family

To realize ourselves as alive means to realize ourselves as family. Totally englobed and absolutely unique. Free to act yet bound by dreadful family ties that require reciprocity, if only to continue breathing, in, toward myself, and out, toward the other. Beauty entails its own ethics. Although the experience of joy and emotional ascent associated with beauty elevates the self, at the same time it points in the opposite direction. Ascent comes through connection, and connection warrants a certain attitude. We can exist only if we don't put our ego in the center, because the skin is always shared. Where mine opens up, yours starts. Where my epidermis blossoms, it meets the breath of the world, which is the faint presence of every being's skin. Feeling the lichen's skin against mine means realizing that I am myself an act of relating, not a separate individual, distinct from other objects. Feeling this skin requests that I do my part to make relating possible.

In the experience of beauty, we feel that we are family. We realize that we are child and parent to what radiates outward, to what calls us and mysteriously already knows us. It is flesh from our flesh, be it as seemingly distant as the colored spheres on a weathered roof, or seemingly as close as the microscopic ridges on a tender finger that touches our palm. Experiencing beauty means to recognize family and to feel welcomed into connection. Only if we forsake it by putting a wall between humans and the rest of living matter does the realization of these ties result in "admiration and grief."

One of the most profound effects of encountering beauty is the impulse to radiate back—the pull to strive to express in words, music, or shape what has excited us by letting us know what we are part of. The experience of beauty incites us to give back by giving away something of ourselves, what Lewis Hyde calls the "labor of gratitude."[1]

Undergoing beauty is therefore a profoundly social process. If we are blessed with beauty, we feel that we owe something. We are in debt to the forces that are continuously creating this cosmos. What is beautiful can be realized only if we reciprocate with our own acts of beauty. Giving back beauty by creating beauty is what drives many artists. Giving back aliveness for having been enlivened is at the center of animist rituals. Both are social gestures in which a person—human or nonhuman—who has been kind to us is treated with kindness.

We can now understand beauty better: It is not the experience of an abstract principle or the glimpse of an ideal world. It is the encounter of another person that shares the desire of the cosmos to be in connection with us. Beauty is a meeting that has gone well, and we wish to give thanks for it by enabling more fecund encounters. Undergoing beauty is a social emotion because the cosmos we are embodying in our flesh is a process of intersubjectivity, of mutual breathing.

Being welcomed by family invites us to respond and to reciprocate. What is required—for our own sense of balance, for the well-being of the person we just met, for the fecundity of our shared cosmic body—might be as simple as saying thanks for a blessing received. We can say thanks in many ways. One way is to politely ask and, if allowed, give a caress with the fingertips. Feel the other's skin and how it feels ours. Let the lichens feel how vulnerable and open your flesh is, and sense how patient and enduring the lichen's is. Feel, and let feel, how in meeting both become one and many.

NOTES

1. Lewis Hyde, *The Gift: Creativity and the Artist in the Modern World* (New York: Vintage, 1979), 249.

AVOESIS: TEN THESES ON KINSHIP WITH BIRDS

Liam Heneghan

There's a form of quietness, that is not precisely silence, characterized by an absence of noise or βοή (*voe*) in Greek, a word that might also translate as "clamor" or "din." This auditory lull asserts a benevolent presence; I call it *avoesis* (that is, the absence of *voe* or noise).[1]

1. . . . in Mr. Joyce's first thunderword (one of ten found in *Finnegans Wake* [1939]): *bababadalgharaghtakamminarronnkon-nbronntonnerronntuonnthunntrovarrhonawnskawntoohoohoor-denenthurnuk*—which, by the way, is not as unpronounceable as you may assume (try it!)—one recognizes an onomatopoeial, Jovian echo of the grumble, rumble, roll, bang, blare, shriek, call and cry, the squall and bawl, of a loquacious sky first heard a century or two after the Great Flood and that according to Giambattista Vico (1668-1744), the Neapolitan philosopher, such a sound terrified those giants descended from Noah's son Shem (who had set them wandering through the great forest of the earth for centuries, until, at last, their facility for speech had atrophied), some of whom were mid-copulation when they were roused by the roaring sky; thus the thunderstruck giants took note of the heavens, which had been mute since diluvial times, and fled into sheltering caves where the shock of the thunder-event delivered a sensory topos that founded civilization, led to the invention of Jove (or Joves), and provided an impetus for the discovery of language instigated in imitation of the storm, just as we "still find children happily expressing themselves . . ." and which thunder, in turn, serves in *Finnegans Wake* to trouble the novel's dream language and thereby record Joyce's debt to Vico whose name is commemorated five times in *Finnegans Wake* (mostly in reference to Dublin's Vico Road), but whose circular conception of history provides a structure for the difficult novel, and is thus echoed . . .[2]

Thesis 1: It behooves you to mimic the sky.

2. In her essay "On Some Lines in the Venerable Bede," Marguerite Yourcenar (1903–1987) reflects on a report that the opinion of a thane in the service of seventh-century Northumbrian king Edwin was solicited concerning the incursion of "a god named Jesus." The thane remarked that human life is like a sparrow flying through a great hall lit by a warm fire while outside the winter storm howls. The sparrow "swiftly sweeps through the hall . . . and after a brief respite . . . he goes back into [winter]."[3] If Christians know more about the storm from which we come and to which we return, the thane's reasoning goes, then why not accept Jesus? The flashing of a bird provides the cinching argument. Birds flit through the enigmatic chambers of Immanuel Kant's *Critique of Judgment* (1790) on several occasions. (For Kant a bird is his "dear little creature," whose song "tells of joyousness and contentment.")[4] Birds roost mainly in a section entitled "Of the Intellectual Interest in the Beautiful." Twice Kant imagines a seeker after natural beauty being fooled by an imitation of a bird. A roguish lad mimics a nightingale: once discovered, can anyone still tolerate that song? The good-souled lover of the free beauty of nature is disturbed by mimicry, for contemplating nature brings to mind the possibility of a creator. The thane glanced about the gilded great hall: he imagined the darting bird and the great storm. He elects a new god.

Thesis 2: The morally good is suspicious of mimicry.

3. On the origin of music: Forty days after the flood, Noah sent out a raven that flew back and forth across the waters. Then, time and again, he released a dove until the dove returned with an olive leaf in her beak, signifying the receding of the flood. Set free on a final flight, the dove did not return. As animals, including humans, swarmed the dry land, both raven and dove alighted in the forested lands of Mesha. Living there for many years, the birds attempted to converse. Whereas the raven could gurgle, grate, croak, and utter shrill calls, the dove merely cooed, although she did so in long and short bursts of song. Neither bird could interpret the conversation of the other. Now it so happened that a young man entered the forest there to live in solitude. After a long silence, he finally spoke aloud, but his voice seemed rebarbative to his own ears. So he imitated the birds, and the dove and the raven understood his loneliness. In understanding this man, they understood each other also. One day the man heard a call that was neither the raven's nor the dove's but a blending of the two. He responded in kind. It so happened that a young woman had entered the forest, and in her forlornness, she too imitated the birds. And thus the man and woman sang to each other with the voices of birds. Theirs was the first music.

   Thesis 3: Music echoes the birds.

4. The following tracks incorporate recordings of birdsong: "Grantchester Meadows" (Pink Floyd); "An Endless Sky of Honey" (Kate Bush); "Blackbird" (The Beatles); "A Salty Dog" (Procol Harum); "Summer's Cauldron" (XTC); "Even in the Quietest Moment" (Supertramp); "Nightingale" (Roxy Music); "Music on My Teeth" (DJ Koze, José González); "Panic in Babylon" (the Brian Jonestown Massacre); "Poem of Change" (Pauline Oliveros); "Picasso" (Michael Head & the Red Elastic Band); "Faultlines" (Karine Polwart, Pippa Murphy); "I Know Why the Caged Bird Sings" (Buckshot LaFonque); "By Your Side" (CocoRosie); "Jâm-e Nârenji" (Mohammad Sâdeq Bolbol, Rahim Khushnawaz, Gada Mohammad, Naim Khushnawaz); "Lovin' You" (Minnie Riperton); "Due West" (Kelsy Lu); "The Killing Moon" (Nouvelle Vague, Mélanie Pain); "Cough Drop" (the O'My's); "One for Jo (Alternate Version)" (Bert Jansch); "Curlews" (Grasscut); "Ask Me No Questions" (Bridget St. John); "Genetic World" (Telepopmusik, Soda Pop); "The Tan Yard Slide" (Sam Lee); "Shark Ridden Waters" (Gruff Rhys); "Paris 1919" (John Cale); "Cirrus Minor" (Pink Floyd); "Summer Teeth" (Wilco); "Yawny at the Apocalypse" (Andrew Bird); "The Park" (Feist); "End of the Season" (the Kinks); "Moving Further Away" (the Horrors); "Dog" (Nat Johnson); "As the Dawn Breaks" (Richard Hawley); "Harvest Moon" (Cassandra Wilson); "Clowns" (Goldfrapp); "Crosswind" (J. Tilman); "KINDRED I" (Kelsey Lu); "Koop Island Blues" (Koop, Ana Brun); "Unknown Caller" (U2); "The Lost Words Blessing" (Jim Molyneux and others); "In Maidjan" (Heilung); "Aerial Tal" (Kate Bush); "Small Hours" (John Martyn); and "My Country Home" (Neil Young, Promise of the Real). One track contains mimicry.

Thesis 4: Birdsong adds to all music genres.

5. “Very pretty is the idea that we derived our music from the birds, from the wind sighing through the reeds, from the babbling brooks,” conceded Henry C. Lunn, onetime editor of that venerable journal of classical music, the *Musical Times*, in his 1866 editorial, and yet Lunn sternly concluded that such a notion “is poetry not history.”[5] The musings of Giambattista Vico and tales of thunder-stunned giants notwithstanding, we will hardly be convinced that the origins of music can be found in mimesis. Despite Lunn’s admonishment, there are some clear similarities between birdsong and the melodic structures of music that call for explanation. The neuroscientist Adam Tierney and his colleagues speculate that the commonalities in melodic shape between birdsong and human song (for example, a tendency for songs to terminate on a noticeably sustained final note) reside in motor constraints on sound production shared by these evolutionary disjunctive animals. Both birds and humans modulate pressured breaths to vibrate membranes (vocal chords in humans; sound-generating labia in the syrinxes of birds) to produce sound. Musicians among the frogs and stridulating insects face constraints but of a different kind. The nightingale and the eventide *contralti* use their instruments as efficiently as they can. Let us update Mr. Lunn: “Very pretty is the idea that the evolutionary origins of our music is shared with the birds; this is both poetry and phylogenetic history.”

Thesis 5: Though they diverged evolutionarily 250 million years ago, neither birds nor humans can resist a final long note.

6. An evening in Evanston, Illinois (August 21, 2019): Raindrops falling from roofs to pavement below with small detonations; a single sparrow calls; car tires hiss on a road in the distance; a sparrow calls over and over; air conditioners hum; cars plash through puddles; moments now with birds; insects remaining silent; bells clang from the Union Pacific North Metra train passing Main Street Station; sparrows close now: really meaning business; susurrus of cars upon the road; the *drip, drip,* from the elm; a bird calls at high frequency, a descending note; the clunk of a fridge door; the snap of a porch door; Angelus bells ring, two in competition, occasionally overlapping; cicadas precede, endure, outlast the bells; a robin's *chackchack;* a juvenile lets out a single note, flies to elm; quieter now, a softer light; no raindrops now, a single bark and the dog falls quiet; cicadas roar, like competing choruses, like choirs in Mahler's Symphony No. 8; sparrows in small clamor; a plane hums overhead; insect chorus out of phase but precisely so; a second plane; a train; a car brake squeaks, sounding like what I remember of bats (when I could still hear them); soft now the insects call, strumming softy; soft the failing light; a footfall upon a neighbor's stairs, jangle of keys, slap of a screen door; cars in singletons; a few cicadas purr, rallying the troops; the elm falls silent; soft the call from the quaking aspen.

Thesis 6: Listen for music; you'll find it.

7. Summertime in Chicago—with its pleasing lake breezes, boisterous street festivals, music in the parks, the pyrotechnics of its storms, its frolicsome beaches, the inebriated comradery of the beer gardens, and cryptic cicadas, now loud, now delicately murmuring in the trees—is so relentlessly beautiful that it induces a peculiar form of forgetfulness: you abandon your plans, carefully wrought the winter before, to leave the city. And then, once again, winter sets in. But for all of its rigors, this season offers the prospect of walking into relatively depeopled landscapes: the parks deserted, the beaches left to their own devices, and the cemeteries, crisp underfoot, empty of all but their permanent and silent denizens. Can one, then, in the frigid months, escape urban noise in the city's open spaces? Is this the season's consolation? In the winter of 2019, with students Bailey Didier, Matthew Rosson, and Ashlyn Royce (joined later by Angela Stenberg), we recorded soundscapes in the city's wintry landscapes. Inclement mornings spent at a bird sanctuary, a restored pond near a busy intersection, and in a gloriously, treeful, decidedly baleful, suburban cemetery. *At no site did silence prevail for more than one minute.* Anthrophonic sounds intruded one-third of all the time we recorded that winter: trains, planes, automobiles. But there were moments, never more than thirty consecutive seconds, where the birds burbled their winter sounds and the wind blew softly in the naked trees.

Thesis 7: Summer and the living is easy; winter and the living is noisy.

8. If a tree falls in a forest and no one is around to hear it, does it make a noise? This is more easily answered than the question as typically formulated, with sound. Unlike sound—which is defined independently of a listener—noise, inarguably, is subjective. Noise is unwanted energy in an electronic system, interruption in communicative exchange, a lack of musical agreeability, a signal without useful information, a screaming of "timber!" in the hushed forest, the crashing as the tree thumps the ground. Noise is disruptive: it is sound that the hearer considers, discordant or attention grabbing. Noise distracts its perceiver from the task at hand. The history of noise began when a discordant sound first disrupted the attention of an organism first capable of being deflected from a purpose. This distinction between sound and noise has no greater champion than Edgar Allan Poe. Listen in on "The Tell-Tale Heart" (1843): the same sonic event transitions from sound to noise. The eponymous (dead) heart intrudes first as "a low, dull, quick sound, such as a watch makes when enveloped in cotton," but it progresses until "so strange a noise as this excited me to uncontrollable terror." Our murderer comes to a devastating conclusion: "I found that the noise was not within my ears." The pattern is repeated—without the use of either word—in "The Raven" (1845): the tapping, monosyllabic corvid drives the narrator to distraction.[6]

Thesis 8: Tree's fall, raven's tap, heart's pound: it may drive you mad.

9. One cannot command birds to sing or legislate for wind to sigh through autumnal leaves. Yet in the fourteenth-century text *The Little Flowers of St. Francis,* we learn that the saint exhorted a flock of birds as follows: "Guard yourselves, therefore, my sisters the birds, from the sin of ingratitude and be ye ever mindful to give praise to God." The birds reverently bowed their heads to the ground, and expressing their delight in the saint's words, there were great outpourings of birdsong. But no matter the commandments of saints, birdsong arises solely from the cryptic volition of birds: mimicry is no substitute. The pleasing sounds of nature can restore us; there is now a skein of research investigating healthful urban soundscapes. But can urban landscapes be managed to allow those enchanting sounds to provide their mental succor? The regulation of urban soundscapes historically has been to mitigate sounds in the highest register with noise control technology and civic work ordinances. Yet a high-quality sonic environment emerges not from abatement alone. Or from utter silence, the deliciousness of a sensory-deprivation tank notwithstanding. No, the absence of disruptive noise and the presence of pleasing sound—a combination I call *avoesis*—occurs when people noise is minimized and natural spaces cultivated. Where can this occur? On beaches, on tree-lined streets, in quiet parks and gardens: there, the birds sing not because we command it, but because, finally, we allow it.

Thesis 9: To guard against urban din, design with birdsong in mind.

10. Outside the Victorian greenhouses of Chicago's Lincoln Park Conservatory one spring morning years ago, an elfin man played tunes upon an Irish tin whistle. This wind instrument consists of a cylindrical pipe with a whistle attached to one end. Notes are produced by blowing while shifting fingers in patterns across the six holes of the instrument's shaft. The effect is unexpectedly complex. The elf played a nickel instrument in the key of E flat. What his playing lacked in finesse, it made up for in gusto. I paused—how often does one hear such music *outdoors* in Chicago?—and, gesturing toward him, I raised my camera. He flinched and, turning his back, withdrew from the frame. I lowered my lens. When he returned, I apologized. "You *never* can tell what a camera takes away," he insisted. I asked his name, and he reported that he was "presently between names," a state of affairs I'd never encountered before. The elfin man was friendly now, though. "Why," I asked, "do you play the E flat rather than the D?" (As D whistles are the more common key.) "Because," he replied, "birds prefer it." Noting my skepticism, he offered to demonstrate. Preparing the instrument, he blew a drizzle of spittle from the whistle and played a lively jig. "Jackson's Mistake," I think it was. Sure enough, sparrows came from their hedges and formed an audience around him. They were silent, listening.

Thesis 10: Birds listen if you know how to talk to them.

NOTES

1. Liam Heneghan, "A Place of Silence," *Aeon*, February 24, 2020, https://aeon.co/essays/why-we-need-an-absence-of-noise-to-hear-anything-important.
2. Giambattista Vico, *The New Science of Giambattista Vico* (1825; New York: Cornell University Press, 1968); James Joyce, *Finnegans Wake* (London: Faber and Faber, 1939). I use the 1975 edition of *Finnegans Wake*.
3. Marguerite Yourcenar, *That Mighty Sculptor, Time*, trans. Walter C. Kaiser Jr. (New York: Farrar, Straus & Giroux, 1993).
4. I used this translation: Immanuel Kant, *The Critique of Judgment* (1790; New York: Prometheus Books, 2000). The translation preferred by contemporary scholars is Immanuel Kant and Paul Guyer, *Critique of the Power of Judgement* (Cambridge: Cambridge University Press, 2008).
5. Henry C. Lunn, "The History of Musical Notation," *Musical Times and Singing Class Circular* 12, no. 278 (1866): 261–63.
6. The version of Poe's stories I use is Edgar Allan Poe, *Poetry and Tales: Vol. 19* (New York: Library of America, 1984).

FOUR TURTLES

Brooke Williams

In Florida, two winters ago, Terry and I wandered along a canal flowing through the J. N. "Ding" Darling National Wildlife Refuge. We were watching birds, but we also saw four turtles.

Three of the turtles were sitting along a large log angling out of the water. Low Turtle sat on the log just above the water level. High Turtle sat midway up the log, with Mid Turtle between High and Low Turtles. The fourth, Shore Turtle, lay in the shade-covered mud on the far side of the canal. The three turtles on the angled log appeared to be sleeping.

Shore Turtle rose from this resting place and slowly moved toward the water. As if the sound of Shore Turtle entering the water were the signal (inaudible from where we stood watching this drama through binoculars), High Turtle rose from the log, stood momentarily, and moved up the angled log, the distance of exactly one turtle length, then lowered himself, resuming his resting position. Mid Turtle then rose, moved up the log, and settled into the position previously occupied by High Turtle. Low Turtle did the same, moving along the angled log, stopping just behind Mid Turtle. This created space for Shore Turtle, who had just arrived at the log to climb out of the water. Now, instead of three turtles asleep on the angled log, there were four.

I'd seen wild turtles before. More than what I saw, I felt something unnamed, yet familiar, pass between those turtles. Thinking about it later, I was able to make sense of it. When I agreed to write about kinship, I had a different story in mind, one I'd already started that I thought could be modified to fit. As the deadline approached and I opened my notebook to begin, I found this very different story poised to be told. I interpreted the behavior we observed as the polite caring of three individuals for a fourth. If this makes me guilty of anthropomorphism—attributing human behaviors and mental states to other species—so be it.

I'm suggesting that we share consciousness with these turtles, which, as a biologist, I'm not allowed to do. Oliver Milman writes in the *Guardian* that anthropomorphizing can be harmful, as "scientists are still keen to draw stark lines of distinction between humans and animals."[1] I don't understand. He's referring to what is found on social media. He quotes Holly Dunsworth, an anthropologist at the University of Rhode Island. "It's almost like the internet was built for anthropomorphizing animals," she said in response to a photo of a kangaroo holding the head of his dying mate. "No one has shown that animals understand dying or where babies come from. We can't say they think that abstractly."

How can anyone say that that kangaroo doesn't understand dying? Why would Dunsworth suggest that holding a dying mate is to think abstractly? Apparently, there are some scientists who can't think their way to knowing that "animals understand dying or where babies come from," and if they limit the possible explanations to what they "think," then this is the result. The kangaroo did not "think" that his mate was dying. But I'd bet that he *knew* that she was dying. These scientists justify putting this behavior through the lens of what we "know," which is limited by what we "think."

Frans B. M. de Waal, a Dutch biologist known for his work on primate behavior, says that anthropomorphism is a product of anthropocentrism. Of course. When we believe that modern humans are the most important element of existence, above all other life, we must refrain from acknowledging that we might possibly share emotions with other life-forms.

I may be attributing those turtles a human mental state, but then back to the question: as humans, do we assume because we have an emotion or "mental" state that it is unique to us as "higher beings"? That "lower" beings cannot possibly share this mental state? These emotional or mental states might not be "human characteristics" but, in fact, may be common to all living beings. What would we call the idea that as living beings we share many emotional or mental states with other living beings? Ecomorphism? (In the same sense that ecocentric extends the status of moral object from human beings, not only to all living things in nature but to abiotic elements such as ecosystems and watersheds, as well.) This is in contrast to anthropocentric, which regards humankind as the central and most important element of existence. Perhaps the real danger of attributing human emotions to animals is assuming that animals are incapable of feeling.

Anthropocentrism—humans as the most important element of life—may be at the root of our current climate crisis. The renowned scientist and environmentalist David Suzuki believes that our current situation results from our having "lost the indigenous knowledge embedded in place." That a future in which our children and grandchildren can thrive requires "a paradigm shift" by which we begin "to see the world as indigenous people see it."[2] There are many Indigenous people who believe in a nonhierarchical, animistic world in which everything natural is interconnected. That their oral traditions often depict animals with humanlike characteristics, such as talking and walking upright, is not considered anthropomorphic but *personhood*, a term that can be applied equally to all life. Shifting the paradigm requires focusing beyond our humanness to a more ecocentric view of the planet. Shifting the paradigm means—and this is the tough one—considering humans as but one species among many, one element of a complex and beautiful kinship system, and *knowing* that the more-than-human world is a source of wisdom and protection.

This is not new. Regardless of our ethnicity, we all have a direct genetic connection to Indigenous knowledge. Although still present in our evolutionary bodies, this biological core is obscured beneath what the great human ecologist Paul Shepard referred to as a "veneer" of civilization, beneath which, he writes, lies

> the human in us who knows the rightness of birth in gentle surroundings, the necessity of a rich nonhuman environment, play at being animals, the discipline of natural history, juvenile tasks with simple tools, the expressive arts of receiving food as a spiritual gift rather than as a product, the cultivation of metaphorical significance of natural phenomena of all kinds, clan membership and small-group life, and the profound claims and liberation of ritual initiation and subsequent stages of adult mentorship. There is a secret person undamaged in every individual aware of the validity of these things, sensitive to their right moments in our lives.[3]

If there is "a secret person undamaged" in each of us, what happened? The damage goes back in time to the Enlightenment, to René Descartes, to "I think, therefore I am." During that time of the great historical shift, we reasoned ourselves into believing that modern humans were not only better than anything else but also in charge.

"Thinking" may or may not be unique to humans. What does seem to be our specialty is the belief that thinking is the only trusted form of knowing. What have we sacrificed by limiting our knowledge to what we can think?

One early spring, years ago, a friend in Idaho took me to see a moose he knew about who had died during the harsh winter. The moose had become food for other creatures, large and small. We sawed through the moose's skull, freeing his antlers. His spongy gray-green brain was so small. "How," I thought, "can this animal manage his huge and powerful body with this small brain?" My only explanation was that the moose's life must depend on many different modes of knowledge and only slightly on thinking. We cleaned away most of the tissue inside the antler part of his skull.

I took it home and placed it on the roof of our garage, where we watched a mother magpie feed three sets of hatchlings on remnants of that moose's brain.

I think.

Darwin wrote "I think" at the top of the page of his notebook on which he diagrammed his theory of natural selection—how he "thought" species evolved.

I think he wrote "I think" because he "knew" that something never before explained was at work, although he wasn't sure how or why. That he needed to know how or why may actually be the uniquely human trait.

In *On the Origin of Species,* Darwin used the phrase "survival of the fittest" to describe the mechanism for natural selection. We've grown up with the idea that this theory of evolution is based on competition. Darwin's *thinking* evolved, however, and later, in *The Descent of Man,* he used examples of cooperation to suggest that it is more likely that working together leads to flourishing and survival rather than to selfishness.

Had Darwin watched the four turtles during his "competition" phase, he may have seen that they were not sleeping but waiting for their turn to mate. Being lowest on the slanted log, Low Turtle was in the best position for mating, and, of the three, the most dominant. But Shore Turtle was actually the ultimate grand male turtle, the strongest turtle in the neighborhood. Although Low Turtle would love to be the one to mate next, he knew that Shore Turtle would fight for and likely win the best mating position on the log. Low Turtle could not risk the embarrassment and loss of stature of being tossed off of the log, knowing that when Shore Turtle had finished mating, his turn would come.

Later, I like to "think," Darwin might have watched the four turtles through the same lens of cooperation and caring that I did.

Regardless, the turtle may not have "thought" any of this, yet somehow all the turtles knew it from some other source of knowledge beyond our ability to think of just what it might be.

Limiting ourselves to what our brains can figure out may be the source of our problems. Or so "I think."

Many times in the presence of nonhuman animals, I've felt the pedestal atop which humans have perched since the Enlightenment being chipped away at, eroding. Years ago, I spent two weeks on a boat, working on a film about killer whales (watching them rise up around us and play and love one another, their conscious breathing, their echolocation). Reading somewhere later that, second only to humans, killer whales are the most intelligent animals on Earth, I thought, "but only if humans make up the test."

Recently I discovered the dragonfly's "third eye"—the ocellus—a small organ between its giant eyes. Because of this phenomenon, dragonflies are essentially flying gyroscopes capable of instantaneous response, making them twice as effective as any other flying predator. Via a complex system of neurons, this third eye, this ocellus, connects directly to the wings, bypassing the brain, eliminating any thought or decision, which would slow down their reactions and undermine hunting prowess.

Many more examples exist of nonhuman animal capabilities of which humans do not have the capacity to dream, let alone understand.

Is it intelligence? A different form of intelligence. We think, and so we call that intelligence. We'd do better to ask, What kind of intelligence does the turtle have? The dragonfly? The killer whale? It is beyond thought; out in a realm off limits to those of us constrained by thought.

That the story of the four turtles insisted on being told is, I believe, my unconscious at work, insisting that therein is something (my "secret person undamaged"?) attempting to bubble up through the darkness. I trust this. Animals appearing in our dreams and otherwise finding their way into our consciousness is older than thought, suggesting that where we end and the other begins is still, as it has always been, indecipherable.

Soon after the four turtles incident, I was talking to students about awe. One of them asked me about the last time I had felt awe, and without thinking, I told them the story of the three turtles moving up the log to make room for the fourth. Awe, according to the psychologist Dachar Keltner, is associated with the shrinking of the "self" in reaction to the enormity of the scene unfolding before you. The "self" here refers to ego, which expands with thought by quieting any deep instinctual knowledge still lodged in us but ignored since we convinced ourselves that we tamed the wild world.[4] As a result, the wild world has lost its importance.

When the three turtles moved to make room for the fourth, I "knew" why. I recognized it with my True Self, that larger, more collective part of myself.[5] Not until later as I "thought" about the experience, did I wonder about other possible explanations. The deep sense I had watching those turtles resulted from my recognition of something very familiar to me but in a different species. This transcended thinking. Thought and reasoning and making sense are luxuries in the wild world, unnecessary. Those turtles did not "think" about caring about one another. They simply cared about one another as if their greater lives depended on it.

The real sin is not that we project our emotions and feelings onto other species but that we assume that other species don't have emotions and feelings that we might be lucky enough to share.

NOTES

1. Oliver Milman, "Anthropomorphism: How Much Humans and Animals Share Is Still Contested," *The Guardian,* January 15, 2016, https://www.theguardian.com/science/2016/jan/15/anthropomorphism-danger-humans-animals-science.
2. Luke Briscoe, "Dr. David Suzuki Reveals Seven Core Learnings from Indigenous Peoples," *Indigenous Peoples Major Group for Sustainable Development,* https://indigenouspeoples-sdg.org/index.php/english/all-global-news/815-dr-david-suzuki-reveals-seven-core-learnings-from-indigenous-peoples.
3. Paul Shepard, *Nature and Madness* (Athens: University of Georgia Press, 1998), 130.
4. Summer Allen, "What Awe Looks Like in the Brain," *Greater Good Magazine,* October 18, 2019, https://greatergood.berkeley.edu/article/item/what_awe_looks_like_in_the_brain.
5. As I understand it, "True Self," a Taoist, Buddhist, and Jungian concept involving the universal self—beyond all identity, to which we all aspire. Alternatively, the "ego," which is the small self, can expand creating noise enough to drown out the larger, universal self.

CHARMED

Susan Richardson

She calls him from a thousand miles distance—
sends forth an invisible cord
from her edge of cliff

to his edge of existence.
She calls him in autumn storms,
in summer stillness,

grooves a new migration route,
moons him towards her
for tide after tide.

Some claim, with disdain,
that she practiced with sprats; some beg
her to tame their cats

and mend their lame horses.
Others whisper that she's aiming
to become one, and hide

from her gaze, for upgrading
to rorqual takes whole shores of sorcery
compared with subsiding into seal.

She knows when he's close
from the tingle of krill on her tongue,
the pulse of his infrasonic hum

in her thighs, the unpleating
of her throat as she hazards a smile.
Though she's tiring, awed by his size,

she relies on no lyre to draw
him; no piping raises him, swaying,
from the waves. And when he finally arrives,

she defies the precipice, leans further,
further, further over the side, to tell
him why she fetched him, words stranding

between them like baleen:
Not to contrive more scientific lies,
not for those men rubbing krónur

from their eyes, but to show
that my kind need not be predators.
For no other reason but to see.

FROM *LOGOS* TO *BIOS*: ON KINSHIP WITH A MUSHROOM

Andy Letcher

My kinship is with a mushroom.

I'd love it to be up there, above our house, with the ravens turning victory rolls through the air, or with the buzzards that sing of the wind, or the jackdaws busy in the farmyard next door, cutting deals. But they're skittish, understandably. I can never get close.

No, my kinship is down below, with a fungus, ignored by most, shunned by many: the liberty cap, *Psilocybe semilanceata*. Every autumn it grows with great abundance in the acidic grasslands of the United Kingdom, but to eat or even pick it is quite illegal. Our affair is star-cross'd, but as the old adage goes, love and music need no passports.

For something not much bigger than your little finger, this goblin-hatted toadstool packs quite a punch. Scientific investigation—the work of *logos*, or the rational mind—has revealed it to be stuffed with tryptamine alkaloids: psilocybin, psilocin, baeocystin, potent psychedelics all. If you're bold enough to eat even a dozen, your world will be completely transformed. You'll be thrown headlong into that other realm: *mythos*. There may be ickiness at first, a feeling of nausea or vertigo, like you might faint. When it passes and you can stand again, or bumble about leaden footed, you'll see that someone's turned the dial up on all the colors. There's an ambiance of enchantment as though you've stepped into fairytale. The simplest of acts takes on mythic significance. And lest you take it or yourself too seriously, the experience comes infused with such a wry and ribald levity that you're left with the distinct impression the whole shebang began with a belch, a fart, and tears streaming down the Creator's face.

I was initiated at college, aged nineteen. An older hippie—let's call him Gary—took me picking, one damp autumn morning, in the Peak District. Gary was permanently stoned on squidgy black Moroccan hash, his eyes half-closed like the Buddha. Eschewing ashtrays, he'd dab his spliffs onto his jeans, rubbing the ash in with his middle finger. He was, I suppose, one of the unwashed, but he knew his mushrooms and was, in a sense, my spiritual guide. Even before he'd turned over my pickings, held them to the light, and given me the all clear, I knew already that I'd scored gold. They might just as well have been glowing, or jumping up and down waving a flag. I know it makes little sense to *logos* but I can describe it only as a kind of preinstalled familiarity. I already knew the mushrooms before we'd been formally introduced. The late psychedelic guru Terence McKenna would say of mushrooms that they were a calling.

Thus began my longest-ever relationship. I suppose I've clocked up hundreds of hours, adventures and misadventures in equal measure. Mushrooms on the beach. Mushrooms in the snow. Mushrooms down the pub on New Year's Eve. That time my birthday party was raided, police helicopter and all. Another encounter with a helicopter, this time an army Chinook, that appeared out of nowhere and landed just yards away like a demon from a ring of hell. And once I went to a club to see the band the Shamen—remember them?—and pulled a whitey and spent the entirety of the gig curled up in a ball in the corridor, willing it to end.

It took me a surprisingly long time to realize I was doing it wrong. In my defense, I was simply acting on what I'd been told by Gary and his peers: take mushrooms at parties and gigs and festivals, free your mind, and stick it to the man. I don't really subscribe to the notion of "bad trips" because if you're doing it to learn something about yourself and the world—and, my naïveté aside,

I was—it's all data. But gradually I recognized there was a pattern to my having a horrendous time. It all went bad when I stopped listening.

In an animist world of kinship, perhaps what is most required is that we listen. Yet ours is an increasingly narcissistic culture in which instant communication has deafened us to one another. We shout ever louder to no avail. The selfie says it all: it's a visible expression of a much deeper and more worrying tendency of retreat from the world. For, contrary to popular wisdom, we are not "disconnected from nature"—how could that even be possible? No, we have *chosen* to be parochial, to look inward, to put ourselves first, to deny the other and keep it resolutely at bay. But animism, and the kinship that emerges from such a perspective, asks that we turn our attention outward, to develop an active receptivity, to attend to the ravens and the buzzards and the mushrooms and the ants and the weather and that hill over there, the one the jackdaws call home. All of it. This peopled world has much to say, but you can't pay attention when there's a band playing or a party in swing or the police or the army or an ever-present concern for how this will look on Instagram.

Sometime during the run up to the millennium, I learned to sit with the mushrooms on my own or with close friends, with music round a fire perhaps, but always far enough away from the profanely human not to get distracted. And when I listened, the message was the same. There's pattern and form and meaning everywhere; *life* everywhere, all the way down. Do you see it now? Forget *logos*. Forget *mythos* even. There is only *bios*, the unfathomable weirdness that sings both of those into being and of which we are inextricably a part.

I was of course, by then, profoundly influenced by Terence McKenna, the great champion of the mushroom. Right up until his untimely death in April 2000, McKenna regaled audiences on both sides of the pond with spellbinding, blarney-kissed, improvised raps, wondrous trippers' tales, calls for an "archaic revival" of shamanic styles and practices. He told us that with high doses, taken in silent darkness on an empty stomach and with eyes closed, the mushroom would literally speak, in something akin to what the philosopher Martin Buber meant by an "I" and "Thou" relationship. The mushroom, said McKenna, revealed that it had originated beyond the solar system, sent its spores through space in search of mammalian consciousness, or any such similar, with which to join in symbiosis. In return for our good stewardship it would reward us with gnosis and revelations the likes of which would put our most fantastic science-fiction writers to shame.

Far out, as Gary might have said. Far-fetched, you might add. Such is the problem with psychedelics. The experiences they occasion are anything but ordinary—why else would we take them? Yet for the uninitiated they sound like so much magical thinking, the outpourings of the irresponsibly insane.

And yet there's more than a little truth in what McKenna was saying. In the twenty years since his death, our relationship with the mushroom has spiraled in, ever tighter. We *are* symbionts. So many people, especially creatives, have seen the world through bemushroomed eyes that suspiciously psilocybic sequences appear, with a knowing wink, in very straight movies (*Avatar, Frozen 2, How to Train Your Dragon 3*). Mycologists have identified around two hundred psilocybin-producing mushroom species worldwide and have worked out the chemical pathways by which they do so. We've spread spores hither and yon. And most important of all, thanks to McKenna and his brother Dennis, we can now cultivate magic mushrooms in our own kitchens or on an industrial scale. Using spores brought back from the Amazon, and through patient observation, experiment, and trial and error, the brothers worked out how to grow *Psilocybe cubensis* on grains of rye. McKenna would later say that the qualities you needed to grow the mushroom—punctuality, cleanliness, attention to detail and so on—were exactly the qualities you needed to take the mushroom. Active receptivity at every stage. Theirs became an extended intimacy.

Liberty caps are nigh on impossible to cultivate. They eschew captivity, refuse to fruit. So here in the United Kingdom if we want to go local, we have to go picking. Through active receptivity, learned in part from the liberty cap itself, that preinstalled familiarity of mine has strengthened over the years. If nothing else, I have a reputation as an excellent picker. I've learned to sense when the weather is right and the mushrooms are up. I can recognize a potential mushroom field because I feel it as a kind of tug. I know a lot about the liberty caps, their ecology and habits, their chemistry, their pharmacological effects on body and brain, their potential as a cure for depression and the other mental afflictions of our age—*logos* again—but my ability as a picker stems from another kind of knowing, from my immersion in the *bios*. Like an endosymbiont, I've taken the mushroom inside of me. We vibrate to the same frequency. If we're in close proximity, we hum. I just know somehow that they're there.

A friend who is a shaman once saw this vividly. She was performing a healing, some unrelated matter. She used the beats from her deerskin drum as stepping-stones to cross from her soul to mine. Alighting, she wondered, "Who are all these little people?" The penny quickly dropped. "Oh."

I say all this not out of self-aggrandizement but to suggest that this is perhaps a latent ability we all possess and may have simply forgotten. In *Being a Beast*, the nature writer Charles Foster describes how his son Tom, when young, had an unerring knack of turning over stones to find toads. In the Amazon, would-be shamans undertake extended and very restricted diets whereby they eat certain plants to know their powers and healing qualities by literally taking them inside. Perhaps this other knowing is how those whirling flocks of jackdaws outside my house seem instinctively to anticipate where to find breakfast. They go differently every day but always at once and seemingly as one mind. In southern Africa, I've heard they call the vulture the "divining bird" because it just *knows* where carrion is to be found.

So then, when I went picking, and in honor of this intimacy, I'd sing to the mushrooms, as I might to a child. I'd say prayers, make offerings, leave as many as I took, giving thanks as I went. If I'm agnostic about McKenna's fantastic extraterrestrial futures, I still resonate with his notion of the archaic revival and his implicit animism. The little rituals I created seemed the respectful thing to do.

I'm in my early fifties now, father to three young children, short of time and energy, and increasingly adverse to risk. I long for sleep, not shenanigans. It's why I've put all this in the past tense. But I did have a mushroom trip just a few years back, quite unintentionally.

My wife was uncomfortably pregnant and away with family, so I spent the day hiking on Dartmoor, the high moor, up where Bronze Age ancestors built a great arc of miniature stone circles, a ring of rings. A holy place once, I suppose, but desolate now, boggy and bleak, the forest cut, the land burned and grazed for thousands of years. As I crested the hill that borders this wide expanse, I stopped for no reason. I turned to my left and twenty paces away were two liberty caps, ringing like the tines of a tuning fork. It felt churlish to refuse. I said my thanks and ate them. They tasted of greasy earth.

Two mushrooms is a trivial dose by any measure. By rights I should have felt little or nothing. Maybe they were very fresh, or maybe the hinges are so well greased by now that the door swings wide with the slightest push. Either way, they were unexpectedly strong. I stood and watched the wind move across the ever-brightening heather, as if it were being swirled by giant hands, then struck out for a lone tree half a mile away. The rain came in, the sheep path dwindled and with leaden feet I struggled to traverse tussocks that were thigh-high and threatening to twist an ankle or snap a bone. My quest seemed foolhardy, doomed. I considered turning back but somehow, soggy wet and out of breath, I made it. The tree stood right next to a gurgling brook of water the color of amber. It felt like sanctuary.

This solitary oak had somehow evaded the grazing sheep to eke out a lonely life in this remote place. It was stunted and hunched against the biting wind, a gesture of grief. I sat with my back against its trunk, listening to the running water, wondering who ever came here. I don't know why, but I looked up, and there was my answer. Directly above me, almost within reach and wedged into the crook of two branches, was a raven's nest. Empty for now, as twisted and black as hammered steel, it hung there like a vision of the cauldron from Celtic myth. I felt as if a curtain had been drawn back and I was being rewarded with a glimpse of something private, sacred even, not for ordinary eyes. In some roundabout, looping way, so typical of *mythos,* the mushrooms had answered my craving for intimacy with the raven. I gasped. I may have even wept. I took it as the blessing it was meant.

My kinship is with a mushroom and I shall go to the grave grateful for having seen the world through its eyes.

CLOUDS OF WITNESSES

Manon Voice

Somewhere in the world
A child is playing outdoors again
Flying a kite made from mylar
Whipping against the back of the wind
Held to earth by a line.

Arms as poles of gravity
The body not quite a bird
The narrower limits of the eye
Stir the first of many magicks.

The sky is no longer a sky
The name we have given it disappears
In a fugue of delight.
The cascade of thick-bodied cumulus
Shapeshifts into a phantasmagoria of animation.

The child, pried open with curiosity
Projects new inventions on the sprawling blue canvas
Laughing forth a thread of strange names
She bestows upon her novel caricatures.

When the child tries to mouth this mystery
It will disembogue in an avalanche of questions
Peppered upon adults torn from imagination—
Tethered to debts.
She feels their skin-scales of fear,

And turns inward and outward
Traversing a mobius strip of
Googled information fused with her own
Astonishment and the fire of wild interrogations.

She will learn that kites
Were first developed in ancient China
And were probably used for signaling at a distance.
Inside, her house is in quarantine,
A cloistered modern settlement
Of screens.

Lately, she has heard the term *distance*
As many times as she has heard the word *China*.

But now these terms transmute into her fascination with kites.
Her mind has rippled out like the wide sea of the sky.
Her lungs have the breath of her new neighbors of clouds
Who gaily etch her future.

Studious in beauty, she has become a lover of simple things.
She will also learn to create new ones.
New ones,
That lift, that fly, that transport,
That help humans be held to earth by a line.

CROSSING THE FENCE

Brooke Hecht

What are the consequences of knowing little to nothing of your ancestors? Or your ancestral lands? What does it mean when the stories of these closest of kin are but a wisp of memory—or not considered much at all?

Talk of ancestors did not feature strongly in my family upbringing. The imparting of explicit ancestral knowledge was absent, unless you counted the rules held by Emily Post, which seemed to have captured the deep respect of at least a generation or two. Even my parents' immediate families were largely distant—or absent—from our lives. "You choose your family; we create our traditions," my mother often said. She and my father gathered their own circle of friends around us who became our aunties and uncles and with whom we celebrated most "family" holidays. And I took her word for it on choosing family—at least until a fateful shuttle-bus ride to the San Francisco airport after a conference. On that bus ride, I was talking with my colleague John Hausdoerffer about the Center for Humans and Nature's Questions for a Resilient Future—a series of questions paired with essay responses that explore our moral responsibilities to our fellow humans and the greater community of life.

"I have a question for you," John said, "What kind of ancestor do you want to be?" I froze, or maybe I was just holding my breath. I can still see the vivid blues and oranges of the bus's plush coach seats. What I remember most is that it was perhaps the most important question my heart had ever heard.

That conversation turned into a conference and a book.[1] I was rapt by colleagues and authors involved in the project that grew out of this question. Brooke Williams's writing spoke directly to me:

> I didn't think much about ancestors until seeing the movie *Amistad*. This film tells the true story of the takeover of a ship, bound for America, by a group of Africans who were to be sold into slavery upon arrival. There is a scene that I've not forgotten during the decades that have passed since seeing that movie. The Africans are on trial. One of them, Cinque, is about to testify. He is speaking to his lawyer, John Quincy Adams[:]
>
> > Cinque: *We won't be going in there alone.*
> >
> > John Quincy Adams: *Alone? Indeed not. We have right at our side. We have righteousness at our side. We have Mr. Baldwin over there.*
> >
> > Cinque: *I meant my ancestors. I will call into the past, far back to the beginning of time, and beg them to come and help me at the judgment. I will reach back and draw them into me. And they must come, for at this moment, I am the whole reason they have existed at all.*
>
> I will never forget the feeling of wonder I had, imagining that all of my ancestors might be out there somewhere in other dimensions, in other worlds, waiting for me to "call into the past" for help.[2]

That thought had never occurred to me either. And now where do I begin, if I were to call into the deep past, to draw my ancestral kin to me? I do have notions of my parents' and my grandparents' ancestral legacies—many of them beautiful. But they have given me painful legacies, as well, far too close in time to have any chance of being silenced or buried. There are the countless tons of precious metals that were wrested from across the Earth by Newmont Mining Corporation and Western Mining Corporation at the behest of my father. And the atom bombs that fell from the sky, sent from the secret city of Los Alamos, where my grandfather was a chemist on the Manhattan Project, his scientific mind serving the terrifying creations of Little Boy and Fat Man. I carry these truths in my bones.

In light of these stories, it seems important to consider which ancestors I might call upon—and for what purpose? It also seems important to consider which family stories I need to tell my descendants so that *they* might know which ancestors to call upon—and for what purpose.

Here is one story I will pass along to my descendants as far along the family lines as I am able. The gist of the story is this: When my mother was four years old, she undertook a great escape from Los Alamos over the play yard fence. As she remembers it, "wild nature" on the other side of the fence called out to her. She was not tall enough to launch herself over the fence, but—in her benevolence—she was determined to free as many of her other playmates as she could. She lifted them across the fence, and according to plan, they ran away into the canyon. Search parties were sent. That the children were eventually tracked down by the authorities does not diminish her feat.

The children were told nothing of the task at hand in the secret city. Apparently, even my grandmother was not told the truth of what was happening. But children hear whispers. Perhaps my mother overheard my grandfather when he told my grandmother that if she ever saw smoke coming from any of the laboratory buildings, she was to put the children in the car and drive as fast and as far away from that place as she could. Perhaps my mother did not hear those whispers. But a heart knows. Deep down, her heart must have known on which side of the fence the land was enslaved to a terrible task and on which side of the fence the land called out in support of flourishing life.

My mother was admonished: "Don't you understand the danger of the mountain lions and rattlesnakes lurking beyond those fences?" My take is that indeed she did. Given the choice, I too would have rather taken my chances with the mountain lions. Maybe it's genetic.

My grandchildren, I am telling you this story so that you will trust the heart you brought with you into this world. So that you will question authority and what that authority purports to be serving. Find the fence crossing. Look to the land that supports life—so that you might learn from that land how to live in reciprocal relationship with life—with land and water, air and stone, creature and soil.

At the 2017 conference "What Kind of Ancestor Do You Want to Be?" our collaborators and meeting organizers, Black Mountain Circle, gave all participants an assignment: Bring the memory of an ancestor to the conference. Create a brief story or poem of seventy-five words or less about them on a single page; add a photo and decorate the page if you wish. Tell us their names, where they lived, and their relationship to their landscape. If you know the kind of impact the land had on them or they had on the land, include that, too. How were you influenced by their heritage?

Not knowing any family stories farther back than Los Alamos, I asked my father for his help. Fortunately enough for my assignment, my father was investing significant amounts of his retirement time into identifying our ancestral lineage, our ancestors' whereabouts, and as many stories of their lives as he could. "Cliché! A predictable hobby for retirement," quipped some of my social worker aunties. But what if this hobby was the result of our ancestors reaching forward to draw my father to them? That certainly would not be so banal.

My question for my father: which of our ancestors had deep relationships with the land, such that they might have understood the living world as part of their circle of kin? As long as I can remember, my longing has been to understand and connect with the world of plants, mountains, and waters. I asked my father to help me find stories of ancestors who might have had such inclinations.

With my father's assistance, I carried these words to the conference about my ancestors on my mother's side: "My sixth great-grandfather, Stephen Holston (Holstein), was the first European settler (c. 1746) on the middle fork of a major tributary of the Tennessee River, known today as the Holston River. Both a Quaker and noted longhunter, Stephen was close friends with the Chickasaw at the Chickasaw Bluffs (near what is now Memphis). When the Spanish attacked the settlement where Holston lived, eighty refugees survived by escaping to the Chickasaw Bluffs with the Holstons under the protection of the Chickasaw. Because of the close relationship between Stephen and the Chickasaw, our Holston ancestors were known as 'White Indians.'"

A longhunter. He would have had a deep understanding of the land. He would have known which plants were poisonous and which were medicine; he would have needed that knowledge to survive. Wikipedia's entry for longhunter begins as follows:

> A longhunter (or long hunter) was an 18th-century explorer and hunter who made expeditions into the American frontier wilderness for as much as six months at a time. . . . Most long hunts started in the Holston River Valley near Chilhowie, Virginia. The hunters came from there and the adjacent valley of the Clinch River, where they were land owners or residents. The parties of two or three men (and rarely more) usually started their hunts in October and ended toward the end of March or early in April.[3]

Longhunters were master naturalists; some would say they were scientists and explorers. But these are the loftier descriptions. Let us also name them for their effect on the first peoples and the land: the infantry of colonization, the commodifiers of skins and pelts, those who widened the paths for the settlers of European descent who streamed in behind them. Let us name my sixth great-grandfather as among the first in this infantry, opening the doorway to longhunters like Daniel Boone and the white settlers who followed—those who came to the Holston River Valley after he left that land, and his name, behind him.

I have only slivers of stories about Stephen Holston. There is a record from September 1753, from a court appearance that is, for me, an important glimpse into his life. His wife, my sixth great-grandmother, was named Judith. Or maybe it was Tudah, or Judah—or perhaps it was Isabelle? The records are not as consistent with her given name as they are with Stephen's (although their family name is variably spelled Holston, Holsten, or Holstein). The record is equally hazy on who my grandmother's parents' were. (For simplicity, I'll call my sixth great-grandmother simply "grandmother.") It pains me that even though my grandmother is at the center of the story recounted by this court record, her name is not even mentioned, nor is that of her own mother, who is also noted. This is the story:[4]

PETITION OF STEPHEN HOLSTEN
South Carolina

To his Excellency James Glen, Esq. Governor in Cheif and Captain General in and over the Province aforesaid, and to the honorable Members of his Majesty's Council

The Petition of Stephen Holston humbly setting forth that your Petitioner lives and resides with his Family on Little Salludy River, and being about the beginning of August last absent from his House about some private Affairs, [about] 40 Cherokee Indians who had come from Charles Town and confirmed a Peace with the Creeks on there Return to their Nation, they surrounded the Petitioner's House and demanded

Provision. The Petitioner's Wife did accordingly give them Victuals and whatever else she could afford that they wanted, but not contented with that, two of them came in and desired to sleep upon the Floor, for which Permission was given them. At the same Time, it now being dark, the Petitioner's Wife retired to her Bed Chamber, and two other white Men Servants to their Room, but no sooner had they been asleep till the said Cherokee Indians surrounded the House, at the same Time firing a great many Guns, and as on each Side of the House there was a Door, the Indians broke open both the Doors, and came forseably into the Petitioner's House, and one of them armed with a Gun endeavoured to forse open the Door of the Room where the Petitioner's Wife lay. She, looking throw a Hole, asked him what he wanted and seeing all of them armed, to save her Life jumped out of the Window with a young Infant in her Arms, and went throw the Woods three Miles to a Neighbour's House where she took Shelter that Night, and in the Morning returning to her House she found that the said Cherokee Indians had robbed her of the cheif Part of her pewther Plates, and dishes, and also her tea Cups and Kettle, and took about 30 Bushels of Corn along with them, and likewise a Mare belonging to the Petitioner, and also another Mare, the Property of his Mother in Law which they took with them, and carried away.

Your Petitioner therefore most humbly begs your Excellency and Honours to procure Restitution of the Damage sustained by these Cherokees, and at the same Time Punishment for the Violence and Outrage committed as to your Excellency and Honours known Wisdom shall seem meet, and your Petitioner as in Duty bound shall ever pray.

Stephen Holsten
Charles Town the 4th Day of September, 1753[5]

A different source provides additional details for the same incident:

> The said Stephen Holstein being called in, and examined touching the several particulars mentioned about the outrages and Robbery of the said Indians and as he declared that he was not at home when the thing happened he mentioned one Charles Wells and James Daley as the whitemen that were then in the Deponents house when the outrage was committed as also William Savage a Constable and Thomas Beamer and James Beamer Cherokee Indian Traders, whereupon it was Resolved that the said (unintelligible word) be summoned to Town to be examined on that matter.[6]

This story of my grandmother stays close to my waking dreams, although I'm not sure which glimmer of her entrances me most completely. Is it that she was able to serve dinner for forty at such short notice? Is it the strength she must have carried to endure long stretches of time as the only woman in a sea of men? Is it that she jumped out a window with an infant in her arms and then ran—for three miles, no less—to safety? Or is it that she did this jumping and running while simultaneously keeping a baby quiet enough to evade notice by the dozens of men surrounding her house? I have a picture of this grandmother in my mind's eye—a young mother, looking through a hole in the wood, making a quick assessment, jumping out a window, quieting the baby (was she nursing while running?), and then fleeing in the dark, all in very short succession.

I am committed to an ongoing search for historical context, insights into the place and time that might shed more light on this story. One point of interest, as noted in the court record, is that my grandmother was visited by a company of Cherokee leaders that evening—those who had been entrusted with the task of securing a peace with the Creeks. What quarrel did these Cherokee leaders have with my grandmother (and apparently not with the constable or two other "whitemen" present)? In addition, James Beamer was in their company. Beamer was a go-between with colonial leadership on behalf of the Cherokee. For example, he had been petitioning the governor of Charles Town for years (and he was eventually successful) to build a fort for the Cherokee. However, still falling short of this goal in the summer of 1753, was Beamer pondering some reticence expressed by the governor as he sat at my grandmother's house having dinner? Perhaps some other shortfalling of business conducted in Charles Town? Would he have held Stephen (or my grandmother) accountable for any of this? Were my grandparents even allied with the colonial leadership? I doubt I will ever know, though their departure from this part of the world in 1776 suggests that they were allied with neither the British nor the soon-to-be Americans. What I do know is that this dinner party ended in gunfire.

Regardless, my grandparents dwelt in this region of South Carolina for about twenty years more. Records show that they left South Carolina in 1776, an interesting year for them to remove themselves from the colonies. When they did move, they went south and west—to Natchez, Mississippi—as noted in the words I brought to the 2017 conference.

From that place and time, I have another story of my grandmother that stays close to my waking dreams. During the Natchez rebellion in 1781, my grandmother's children—and their spouses—fled from Natchez with their Chickasaw allies. By this point, however, my grandmother's running days were over. Stephen was sick; she stayed at his side as the rebellion broke out. This earned her capture by the Spanish, and she (not Stephen!) was sent to New Orleans as a prisoner. The Spanish hoped that taking my grandmother prisoner would compel her sons and their Chickasaw allies to surrender. This did not happen. Instead, James Colbert, a trader who married into the Chickasaw and whose sons were Chickasaw chiefs, wrote to the Spanish for my grandmother's release. He addressed Don Bernardo de Gálvez, the governor:

To the most excellent Señor Commandant in Chief of New Orleans and Governor of Luisana, etc., etc.

Señor: Madame Tudah Holsten having been taken prisoner and sent to New Orleans, where according to all probabilities she remains in that condition, I should be pleased to have her exchanged for five men whom I hold as prisoners in this nation, as soon as possible. If not, I shall deliver the five men to the Indians.

I am the humble servant of your Excellency.
Jaime Colbert, Capitan.[7]

Colbert's letter came to the Spanish by way of the wife of the Spanish lieutenant governor and governor of the "western posts." The wife of the lieutenant colonel was herself kidnapped by a party of Colbert's men, and when she was released, she carried the above letter in her hands.

My grandmother was eventually freed. (It was said that the Spanish released Madame Holston as a result of the respectful treatment Colbert and his men showed the lieutenant colonel's wife while she was their prisoner.) It is unlikely that Stephen survived the period of my grandmother's imprisonment by the Spanish. I don't know that she ever saw Stephen again. What I do know is that in their lifetimes, both saw much. They were edge dwellers, their feet crossing worlds, British and Cherokee, Spanish and Chickasaw. Stephen was born far from the lands of his own ancestors, and his life as a longhunter meant he would have often been among the few people of European descent in any given place. But I have only wisps of stories. How is it that our family has already forgotten so much of what they must have once known?

Today, I reach back and call into the past to these grandparents, Stephen and Tudah, to draw them to me. Grandfather, what was your relationship with the creatures you hunted? Did you consider furs you gathered kin or commodity? My imagination strains to picture the abundance and beauty of the living world beneath your feet and in the skies above you. Grandmother, did you grow or tend medicines? What was your relationship with the plants you might have harvested? Will you guide my hands well today, as I tend relationships with the plant medicines in the land to which I belong?

Grandmother, what was it like to be imprisoned and far from your family? Did you consider trying to compel your sons home to secure your own release as the Spanish hoped? Or did you perhaps carry a deep sense that you could and should withstand imprisonment so that your sons and daughters might remain with the Chickasaw? What commitments to your kin carried you through your imprisonment? How far and wide did your kinship networks extend?

Grandmother, Grandfather, have you been waiting for me to call into the past for help? I call to you now, from the summer of 2020. I am in quasi quarantine, because of the worldwide spread of a novel coronavirus. Breonna Taylor and George Floyd have just been murdered. The land of your birth has long seen the injustices of colonialization, structural racism, and the destruction of life for "goods and services." There has been a pursuit of happiness—not meaning or purpose, but happiness—and this pursuit has come at great expense. Our human and more-than-human kin suffer the injustices of this pursuit, and my heart breaks with sorrows and regrets that stretch back through time. It stretches forward as well, knowing that unchecked climate change and the deep roots of systemic racism promise more suffering to come.

Grandmother, Grandfather, what kind of ancestors did you want to be? Have the past 250 years brought what you had hoped for your descendants? What are your regrets and sorrows? Put the apology in my mouth; help me speak it now. Help me speak my own apologies. And please do not forget to whisper into my ears what your heart knew of the living world—the wisdom you must have carried. Do you have a vision of the ancestor I might become? Today, I too am an edge dweller, between worlds. Please help lift me across the fence—to the place where we humans can stand in reciprocal relationship with life—land and water, air and stone, creature and soil.

NOTES

1. To view the conference themes and participants, see the website https://blackmountaincircle.org/programs/geography-of-hope/. John Hausdoerffer, Brooke Hecht, Melissa Nelson, and Kate Cummings, eds., *What Kind of Ancestor Do You Want to Be?* (Chicago: University of Chicago Press, 2021).
2. Brooke Williams, "Cheddar Man," in *What Kind of Ancestor Do You Want to Be?*, ed. John Hausdoerffer, Brooke Hecht, Melissa Nelson, and Kate Cummings (Chicago: University of Chicago Press, 2021).
3. "Longhunter," Wikipedia, https://en.wikipedia.org/wiki/Longhunter, accessed July 2020, last edited March 20, 2021.
4. Spelling, grammar, and punctuation have been left as they were presented in William L. McDowell Jr., ed., *Colonial Records of South Carolina: Documents Relating to Indian Affairs May 21, 1750–August 7, 1754* (Columbia, MO: State Commercial Printing Co., 1958), 459–60.
5. McDowell.
6. "Re: Stephen Holstein, Judith King: VA, TN, SC, MS," https://www.genealogy.com/forum/surnames/topics/holstein/115/.
7. Louis Houck, *The Spanish Regime in Missouri* (Chicago: R. R. Donnelley & Sons, 1909), 220.

BURNING OUR KIN

Freya Mathews

Midyear in 2019, I was shown a photograph of a reindeer with its antlers on fire, ostensibly taken on location in the Arctic. I am not sure now whether this was indeed a genuine photograph, but it has stood for me ever since as a mythical trope for the era our planet has now entered. Throughout 2019, Earth seemed to be announcing, by way of an unremitting array of apocalyptic portents, the grand, official, gala opening of climate change: climate change was no longer the future threat about which we have for so long been warned. Its full blast was no longer twenty years off, ten years off, five years off. Its full blast was now, and from this point on there would be neither averting nor turning back.

While the new reality was announcing itself via an Old Testament–type lexicon of catastrophe, the global zeitgeist was also swiveling on its axis. In the space of a year, unprecedented movements like Extinction Rebellion and the school climate strikes exploded onto the political scene. There was a mythic quality to these upwellings also, just as there was to the geophysical disturbances. The new zeitgeist, swirling into being, seemed to have formed in its depths an inchoate need, a hole at its heart that it sought to fill, blowing softly and searchingly over the planet till it encountered a figure—a child—who fit the shape of that hole perfectly. It siphoned her up into its myth-space and set her walking on a path as close as we have seen in these secular times to destiny, in order to stare down the powers driving the ecological apocalypse. These powers had likewise found mythic manifestation in the peculiarly cartoonlike figure of Donald Trump—a figure as unexpected yet apposite, and as freshly minted (and named) on the mythic level, as any archetype from ancient lore.[1] In these two figures, the great rival narratives of our era commenced the fight that will be played out from here on, a fight perhaps to the death: on the one hand, the little girl, defenseless but for truth, standing in fearless solidarity with all living beings, great and small; on the other hand, the solitary figure of the male narcissist writ large, swallowing everyone and everything in his path, pumped up balloonlike to global proportions with the air he has sucked from his victims, ready to swallow the entire world if need be to fill his insatiable inner void.[2]

The new zeitgeist also expressed itself in the eruption of a popular discourse of imminent civilizational collapse, well summarized in the title of David Wallace-Well's widely read essay "Uninhabitable Earth."[3] In response to this startling literature, many of us felt impelled to dedicate our own personal resources more fully to the cause, ramping up our activism and repeating our longtime message that a change of worldview was required. Although I shared these responses, they nevertheless struck me as inadequate. The rapacious extractivism of our modern economy was not merely a contingent formation that could be overturned by purely external means—whether political or philosophical. It was surely underpinned by psychic structures in which we ourselves would prove obdurately invested. Take my own case, for example. It was clear to me, after thirty years of helping to articulate an alternative paradigm through the medium of environmental philosophy, that reason in itself did not change people's consciousness to any significant degree. Yet when I asked myself whether I should give up such fruitless philosophizing and try other means, I detected a kind of mute inner resistance that informed me that, no, I would not be giving up philosophizing, whatever the stakes. One could analyze this psychic investment at length, if one wished, in terms of identity constructed around achievement, discipline, production, distinction, currency—all psychic introjections of an economy defined in terms of production, property, competition, and power. Our inmost identity, in other words, is organized according to the same competitive and combative principles that govern our external, economic relationship with the world around us.

So it felt to me that we could not change our modern economy, and hence the course of civilization, without at the same time changing the hidden conformation of the modern self. In books, talks, and articles, we might eloquently speak truth to power, but the message conveyed through our words would be twofold. One part of the message, about the dangers of climate change et cetera, would be overt; but the second part—about writing as a means of achieving voice and distinction—would be subconscious and covert. The covert message, which inwardly reinforces the principles of production, property, competition, and power, has perhaps in the past consistently undermined the overt one, where this might help to explain why rational discourse has so dismally failed to change our deeper attitudes and the economic attitudes that express them.

Such was the state of play *chez moi* till fairly late in 2019. Then something occurred that was unprecedented in my lifetime, in the history of my country, and indeed in the recent annals of evolution. Australia as a continent began to burn. Not the usual seasonal fires, which are bad enough and have for years been worsening. This was something new. It pointed to the truth that the geological era we have entered is not merely an Anthropocene but a veritable Pyrocene.[4] The fires commenced on the east coast in July, in the middle of winter, months earlier than the official start of the fire season. They took hold in forests sucked dry by years of drought, and thereafter they did not relent. For week after week, month after month, they increased in scale and ferocity, burning out not only the flammable sclerophyll forests but the wet rain forests that had evolved over millions of years to withstand fire. By December, vast areas were blazing out of control. The Currowan fire alone burnt out almost five hundred thousand hectares on the south coast of New South Wales and raged for seventy-four days.[5] Throughout these months, Sydney and Canberra, as well as the fire zones themselves, were dim with smoke, rendering atmospheric conditions extremely hazardous for millions of people.[6]

At first, I felt somewhat removed from what was happening. True, the fires seemed larger than usual, but the news coverage was relatively muted, and where I live, in Melbourne, the weather remained unseasonably cool and mild. By the early days of January, however, I, like most Australians, had become gripped by the unfolding spectacle, on screen, of our nation in chaos—of thousands of people stranded amid infernos on the southeast coast; of people being evacuated en masse by military vessels from lurid beaches; of highways choked with the traffic of fleeing holiday makers; of community after community receiving the warning "You are in imminent danger and need to take action now" as towers of flame bore down on their homes.

Over on the south coast, Kangaroo Island, iconic wildlife haven of over four thousand square kilometers and home to a flourishing koala population of fifty thousand, was meanwhile undergoing incineration. Along with the rest of the world, I gazed transfixed with horror at images of blackened koalas, their still-living faces mutilated and scorched. I listened to them cry out in terrified little voices as they scrambled over glowing red ground on bare feet, their fur on fire. My heart cracked as I watched desperately parched koalas, in video after video, approach a human stranger for help, reaching out to hold their benefactor's hand as they drank from the proffered water bottle. The shattered expressions of the little patients, bundled up in bandages and bunny rugs in the backyards and living rooms of saintly wildlife carers, were hard to bear. I remembered my own foray into burnt-out fire fields in earlier years, searching for survivors to deliver to carers. This was also koala country, and each charred body we found told the heartrending tale of the animal's last moments. Particularly memorable was that of a mother koala, still gripping the base of a tree with one paw and the hand of her collapsed child with the other; also that of a koala pressed into a slight excavation in the side of a stream bank, the depth of the excavation too pitifully small to afford protection. Koalas are more vulnerable to fire than most of the larger mammal species, being so slow moving and so dependent for everything—food, shelter, and safety—on the dangerously flammable eucalypt.

There were of course videos and photographs of numerous other animals grievously hurt or charred beyond recognition—kangaroos, wallabies, possums, and wombats, for instance—and commentary on the tragic losses to conservation that had been incurred: six million hectares of threatened species habitat destroyed; the ranges of approximately seventy nationally threatened species reduced by 50 percent; and, the figure that shocked the world, more than a billion mammals, birds, and reptiles killed.[7] But, given the circumstance of the Kangaroo Island fires, and their impact on the last large colony of disease-free koalas in Australia, it was koalas that became the face of the Great Fires to the world.

Eventually, in February, cooler, wetter weather arrived in the fire fields. But we knew that the fire season was not over; perhaps it would never be over. We knew that this was how it was going to be henceforth: that our forests are drying out and that our new climate might no longer be able to sustain forests. Forested areas might revert to the arid shrubland that occupies most of the Australian continent. It is hard to know what conservation might mean under these new conditions, when the painstaking and expensive work of decades is utterly canceled out in a few days or weeks. The epic suffering we have witnessed our wildlife endure is not over; it will be repeated endlessly into the future. Perhaps eventually we shall lose our heritage of wildlife altogether, the wildlife that, as Australians, we take for granted, abuse, trivialize, but clearly also love.

This outpouring of love was for me the most surprising response to the fires. In the past, the deadly toll of bushfires on wildlife has been almost entirely ignored in news coverage and commentary. The impact of fires has always been measured in terms of asset loss (including the loss of so-called livestock) and the loss of human lives. But now the plight of wildlife was being captured on phones and broadcast to the world in real time. Animal terror and torment was no longer backgrounded as a "natural" part of fires but seen for what it is: individual trauma experienced exactly as we would experience it ourselves.

As the world outside looked on, and we were aware of others perceiving us and our wildlife as interchangeably Australian, we realized that that is indeed who we are. We are the Koala people. The Kangaroo, Wallaby, Wombat, Kookaburra, Emu, Cockatoo, Platypus, and Lyrebird people. For a hundred years, non-Indigenous Australians have asked the question, Who are we? What makes us Australian? Was it Gallipoli and mateship? Don Bradman? Australian rules football? Holden Utes? The barbecue? It is hard to credit that we asked this question, when the answer was so obvious from the start. We are Australian because we inhabit the continent of Australia. The fact that we asked the question at all reveals our entire colonial history of erasure and denial, our blind determination to treat this continent—already a federation of innumerable human and other-than-human nations at the time of European invasion—as a blank slate on which to inscribe a new fiction: Australia. A new fiction which could never "take," or get beneath our skin, because all the while the continent itself was getting under our skin, and making us who we were, a people unconsciously shaped by the particular patterns of light and shade, sound and smell, wetness and dryness, form and, especially, life-form, that were unique—so unique—to this particular continent. So we prattled on about our institutions, our history, our sports heroes, our product brands, but as soon as those of us who were born here went overseas, we missed the smell of eucalyptus and the feel of summer on the beach or in the bush and knew that we were Australian.

It is surely because we have been so blind to who we are and have purposely built Australia as a fiction on that founding nullification of the entire variegated, intricately lived-in character of this land that we have from the very start treated so abominably. In doing so, we have not only brought it to its present state—desiccated, its soils exposed, cooked, eroded, leached of nutrient, polluted, blown away, its forest skin ripped off, its remaining wildlife battered, diseased and on the run—but we have betrayed ourselves. Because we are, at the end of the day, despite everything we have done, *of this place, this land*. It is because we are *here*, and not for any other reason, that we are Australian. And this land preexisted us, by temporal orders of magnitude beyond reckoning. We did not make it. It made itself. It was made by Koala, Wallaby, Long-Necked Turtle, Kookaburra, Bettong, and myriad other species. They shaped it, with their feats of ecological engineering, at macro- and microlevels, as it in turn shaped them, their anatomy, their adaptations, their ways of life. It was their country; its character was an expression of their character, as theirs was of its. In more recent eons, the land was also of course shaped by Indigenous nations, who always acknowledged that theirs was a cocreation with that myriad of other species, each a nation in its own right. When Europeans arrived, their children too were born into Koala Country, Pobblebonk Country, Goanna Country, Cockatoo Country, as well as into the many Countries of the First Peoples. Whitefellas might deny this all they liked, but identity at the deepest level is made not by fiat but by ontological realities.

Perhaps, as we watched these places, this continent—to which we belong in our ontological depths—reduced to ash without our consent, under the eyes of an onlooking world, many of us did indeed realize this is who we are. While the rest of the world wept for the helplessness of the adorable koala, we wept because Koala—along with our entire extended faunal family—is not only helpless and adorable but also, beneath the skin, our deepest kin.

So, Australia wept, in a new way. Throughout the ordeal, people maintained their daily deadpan Aussie personas, their down-to-earth lives, but when talk turned to the fires, many of us found ourselves choking up, ambushed by a grief that lay deeper in our being than we could reach with our words.

After existing in this state for a month or two, numb with grief, barely able to speak about the fires, I, for one, suddenly discovered my response. This response was, I realized, personal rather than political: I could simply no longer bear to live in a way that resulted in the incineration of kin that I loved with a love that was deeper than consciousness could reach. For let us not obfuscate: the Great Fires were the direct result of ordinary actions that I and countless others have taken every day of our lives. In the past, I had of course considered my "carbon footprint" and reduced it where I conveniently could, but I had reasoned that to continue my "important work" toward environmental reform, I would need to play the part of a respectable professional. If this meant traveling to international conferences, keeping up middle-class appearances, supporting a busy schedule with all the amenities of our normalized Western affluence, then so be it: it was what any rational cost-benefit analysis required. Now, however, this kind of abstraction from the actual consequences of such habits has come to feel unconscionable. The issue was suddenly no longer abstract but concrete and personal; it wore a face: that of Koala. I simply could not bear to continue burning Koala and all the other animals whose identity was fused at the root with my own.

In one sweep, then, the fires have cleared away all such rationalizations by revealing a depth of significance so terrible it reduces every other measure to triviality. I now have no choice but to disengage materially from my civilization to a degree it had not before been possible even to contemplate. There must be no more air travel; I must give up driving; food must be thoroughly, responsibly sourced; I must rid my life of plastic; any further commodities or clothes I may require must be purchased secondhand. When all this is negotiated, I may again take stock, and consider which further agents of death my manner of living is unleashing on my kin. I am not prescribing this as a new set of abstract moral standards to which all must be held accountable. This is nobody's business but my own. I am accountable directly to the Earth, to this land—to Koala, if you like—and not to anyone else. Whatever consequences such personal reforms entail for my social identity—my identity as a respectable professional and member of society—is no longer the issue. If I emerge from the process shabbier, gap toothed, less professionally available and less presentable, so be it. Professional obscurity is far preferable to living in murderous denial.

Such defection from the shared material practices of one's own society of course in no way detracts from the necessity for intensified political activism but adds a new depth of sincerity and authenticity to that activism. Nor will it by itself entirely root out the psychic underpinnings of the modern economy, categorically premised as that economy is on brutal instrumentalism. But it might at least begin to close the gulf that has hitherto yawned between our words, as ecological thinkers, and our lives, where this gap may have been rendering our words, for all their validity, hollow. In any case, I know that it is the existential step that I myself must take. By taking it, and by hopefully finding companions in the endeavor, I might free my own psyche sufficiently from the dark spell of our present—mutant—civilization to begin to envisage pathways to the future a little more clearly.

In her classic ethnography *Dingo Makes Us Human*, Deborah Bird Rose relays from her Aboriginal teachers a set of Dreaming, or Ancestor stories, about Dingo and Moon.[8] Moon, in its endless waxing and waning, follows a transcendent trajectory that enables it to rise above the suffering and mortality of fleshly life by subjecting the rest of creation to its will, annexing their life to preserve its own in perpetuity. Dingo declines this path and throws in his lot with his perishable, ever-morphing, fleshly kin, trading the prospect of sterile sameness throughout eternity for identity embedded in the teeming regenerativity of creaturely life. He accepts that, as born beings, we belong to one another, not to ourselves. It is they who make us who and what we are. Aboriginal people, according to Rose's storytellers, adopted the path of Dingo as their Law. By entrusting the story to Rose to pass on to the world, these elders implied that whitefellas incline to the Law-negating path of Moon. The Great Fires have now perhaps given us a new iteration of this story: by following the logic of Moon, annexing the life of this continent to preserve our own disembedded individual identities in perpetuity, we are burning up the very thing that anchors us to reality. We are extirpating the greater family of beings born of this continent, to whom we, as Australians, owe our identity, to whom we belong. Moon shines down as indifferently and serenely on a burnt-out continent, with its charred remains of Koala, Kangaroo, Wallaby, and all the rest, as on a continent teeming with multibodied, multiminded, interdependent life. But Moon is not human at all and can certainly never be Australian.

NOTES

1. A mythic anticipation of Greta Thunberg may be found in Hayao Miyazake's 1984 anime movie, *Nausicaa of the Valley of the Winds*.
2. In another of Miyazake's films, *Spirited Away*, the portrait of hungry ghost, No-Face, offers an uncanny mythic anticipation of the insatiably needy figure of Donald Trump.
3. David Wallace-Wells, "The Uninhabitable Earth," *New York Magazine*, July 9, 2017.
4. Stephen J. Pyne, "The Planet Is Burning," *Aeon*, https://aeon.co/essays/the-planet-is-burning-around-us-is-it-time-to-declare-the-pyrocene.
5. "NSW South Coast Fire Finally Out after Burning for 74 Days across 499,621 Hectares," SBS News, February 9, 2020, https://www.sbs.com.au/news/nsw-south-coast-fire-finally-out-after-burning-for-74-days-across-499-621-hectares.
6. *The Guardian Australia* provided superb coverage of the fires throughout their duration. See, e.g., Amy Corderoy and Lisa Cox, "Counting the Cost of Australia's Summer of Dread," *The Guardian*, February 11, 2020, https://www.theguardian.com/environment/ng-interactive/2020/feb/11/counting-the-cost-of-australias-summer-of-dread.
7. Michelle Ward et al., "Six Million Hectares of Threatened Species Habitat Up in Smoke," *The Conversation*, January 18, 2020, https://theconversation.com/six-million-hectares-of-threatened-species-habitat-up-in-smoke-129438. At mid-February, the final figure for the total area burnt was 12.6 million hectares. See fact check on ABC Radio National program, *Science Friction*, February 16, 2020.
8. Deborah Bird Rose, *Dingo Makes Us Human* (Cambridge: Cambridge University Press, 1992).

"MY FATHER IS THE HILL OVER THERE": A CONVERSATION ON THE KINSHIP OF LOSS

Shannon Gibney and John Hausdoerffer

Traditional, Western notions of kinship privilege biology and genetics—particularly in the form of heteronormative, white, middle-class, Christian family groups. In this interview, the writer, critical adoption scholar, and transracial adoptee Shannon Gibney discusses an alternative, nonnormative kinship formation with friend, colleague, and *Kinship* coeditor John Hausdoerffer.

Theorizing "the kinship of loss," or the positive connection transracial adoptees may feel to each other given the shared loss of their birth families and home cultures, Gibney argues that kinship can be less biological and more circumstantial. It can be a matter of identification. It can be a decision.

John Hausdoerffer: What comes to your mind and heart when you hear the word *kinship*?

Shannon Gibney: Kinship is reflection and refraction, seeing yourself reflected outside of yourself, whether that be in nonhuman beings, like animals and plants and the landscape, or in intangible things—things that you can't always hold such as air and water. We even establish kinship with feelings. So kinship is multifaceted at the same time that it's elusive.

As for my experience, I identify as a mixed-Black transracial adoptee, which means that I was adopted by white folks when I was five months old. You know, when we talk about adoption in the Western context, there's a lot of saviorism tied up in it. We talk about it in these really overly positive terms. It's not that adoptees and adoptive parents don't get the boon of a new type of love and new relationships. It's just more complicated than that. It's always more complicated because there's also tremendous loss and trauma that comes about. The "rekinning" process of separating children from certain families and grafting them into others carries so much trauma.

For me, having the embodied experience of living through that identity and those processes, kinship has also always been both fluid and unstable. For example, while my adoptive siblings are my brothers and there are deeply familial ways in which we existed in this house together as this family, there's also a way in which they're my brothers just because the state said so. There's this constructedness that I think people who are in interstitial places, you know, like adoptees in families and in various relationships, get to feel. We uniquely experience the reality that kinship is constructed.

For me, I've always experienced that constructedness as, "Yeah, you are my kin, because basically society decided that you're my kin."

John: I want to honor another anthology that you contributed to, *Ethnicity and Kinship in North American and European Literatures*, edited by Sylvia Schultermandl and Klaus Rieser.[1]

Shannon: Yes, it will be out this March 2021, with Routledge.

John: Great, and I hope our readers will check it out, because you have an essay in there which gets at what you call the kinship of loss. And at one point in your essay, you say this *kinship of loss* can be a "radical antidote to the profound isolation and melancholy that have brought so many of us transracial adoptees into being."[2] Could you talk more about the difference between the kinship of loss and the melancholy of isolation?

Shannon: That's a really profound question. I'll kind of go back a little bit to an example of when my son Boisey was born with a sixth finger. His dad had a sixth finger when he was born, too. They [Boisey's father's parents] opted to have it taken off. It doesn't have any movement; it just sort of hangs there. But people would, like, jump. Even people in his family would jump when they shook his hand. So this is something that was passed down, is my point, on the paternal side. Boisey's dad has this mark on his hand where the sixth finger was, and Boisey had his sixth finger removed, too. So now he has a little mark as well.

In thinking about this kinship of loss, you know, my son was, like, three and he said, "Dad and I both have the mark. We both used to have the sixth finger."

That family connection got me thinking about how through losing something—and it could be in your genetic line like a finger or it could be a shared experience—through this shared experience of losing something, there's going to be some kind of marker for that. And transracial adoptees carry an intangible mark and thus an intangible connection. There's something that happens when transracial adoptees get in a room together and we start talking together. It's a sensibility. It's a sense of humor. There's a lot of people that don't understand adoptee humor, because it's a bit dark. (You have a little bit of adoptive humor, John. You don't know that, but you do.) Transracial adoptees inhabit kind of an absurd social location—to be the only Black person in a white family, in a white community. And that's not to say there's not love there. That's not to say that it wasn't the right thing to do, to adopt me. People always want to push adoptees into binaries, accusing them of saying adoption is bad. No, I'm not saying that. I'm saying that it's complicated.

John: Shannon, even in having to give that caveat to make sure the reader knows that you were loved, while speaking honestly about your experience, reveals the kinship of loss. I mean, anybody who knows your story knows your parents love the heck out of you, but the fact that you still have to outwardly confirm that in an interview shows how deep that kinship of loss goes, right?

Shannon: Yes, that's exactly right.

John: I noticed in your writing that you're really energized when you talk about other transracial adoptee authors. When you quote Sun Yung Shin on "reckoning" with melancholy, when you quote Saidiya Hartman on kinship identity produced by negation, I can just feel how hungry you must have been and then how fed you must've felt when you found those authors.

Shannon: I think the thing about the social identity of being an adoptee, a transracial adoptee, is that it is a structurally isolating position. It's really hard for us to find each other as children, as teenagers, as younger people, and even as older people. I didn't really find other adoptees until my mid- to late twenties. And that was because I moved to Minneapolis and St. Paul, which is "The Land of 10,000 Lakes." Actually, the big joke among adoptee communities is that it's "The Land of 10,000 Korean Adoptees."

And so those folks really politicized me and also helped me exhale. They helped me grapple with stuff I didn't even know I had, to somaticize what was in my body and to feel like exactly what you're saying—that it's OK to voice the pain and the trauma around adoption. Not having to explain things to people is such a relief, right? It's like a group of Black women together.

It doesn't mean that we're all going to link hands and sing "Kumbaya." That's not what I'm saying. Not everyone is in the same place. There are some adoptees who don't really want to talk to other adoptees, don't really want to think about the pain of being given up by their biological families or, you know, even some of the racism that inevitably comes with growing up in America and then in a white-dominant community or white family. I get that. That's fine. What I'm saying is, there's a language of kinship. And a lot of it is unspoken.

John: That tie to language reminds me of how some of the authors in this collection see *kin* as a verb so that you're "kinning" when you find each other through a kinship of loss, through finding a common language, a common humor, as you said. Is it possible that, through kinship of loss, one can emerge out of the melancholy of isolation? Is the sort of kinship you are talking about more powerful because it's nonbiological, because it is (as you say in *Ethnicity and Kinship,* "fictive"), or is there always a melancholy that will be there because it's not biological?

Shannon: I think those are really great questions. The oppressive part of kinship comes from the dominant cultural narratives saying what a family looks like—heterosexual, white middle-class Christian, biological, et cetera. Any family that deviates from *this* is not a real family. And you're not real kin unless you have *that*. Melancholy can come out of such narrow views of kinship.

But I think it is liberatory to engage in these fictive kinship actions. I think embracing those can release some of that melancholy. I'm a Buddhist, and a more radical aspect of this is embracing that there is melancholy and there is pain in just being alive, and some of us have more than others, you know?

My mom is a former therapist, and she says we live in a profoundly grief-averse culture, which creates all kinds of other problems. Because grief is a part of life. With melancholy, it is kind of like beats, I guess, in that melancholy is going to come back around again.

We now know about, for example, how babies connect with their mothers in those first few weeks and days—there's so much that happens there. And on top of that, loss of the first family, loss of cultures, loss of languages—all that stuff. A lot of times, it's so deep that people don't even have language for it. I have one friend who tried to learn Korean again, who came to the United States when she was three and a half. She has not been able to relearn the language. It's because of this psychological break that happened. That's not an uncommon story.

John: Is the psychological break not just from language but also a break from the taste and smell and sound of the family kitchen; or the feel of the soil and the climate from which the family or even the ancestors cultivated the food; or the stream, pond, well, or pump from which they gathered the water? I often wonder if the feel of a certain soil or thickness in the air from a certain part of the world is ancestrally built into our DNA. We know that trauma is stored in our DNA, and passed on, so why not be stored from a break from an ancestral place? To me, that pain and melancholy is real for all in a transient society alienated from place, and I suspect it is pronounced for your community, for your kinship of loss.

Although I could never fully understand, I feel your point about disconnect from the mother in quite a visceral and literal sense. Do you also expand the literal scope of mothering into the more metaphorical but just-as-real mothering capacity of the soil, of communities growing food together, connecting to ancestral instructions on the land and in the kitchen, merging the tastes, the recipes, the laughter of those fields and kitchens and families? It is unsettling how much of a massive severing you are getting at, yet equally inspiring to see how much you are seeking to heal through the kinship of loss.

Shannon: Right. Again, in the Western understanding of adoption, which is the dominant model, we're not allowed to really contend with the enormity of that loss. I have dear friends who are Korean adoptees who have written about a word in Korean called *Han*. As they have explained it to me, *Han* is exactly what you're talking about. It is not just a loss and longing for the mother. It is deep, existential loss for everything that the *motherland* means, everything that it is.

Even if you have all this love and this fear, and positive relationships with your adoptive family and that new community they provide, that loss is not necessarily going to go away. And even if you have this connection through the kinship of loss with these other transracial adoptees, the melancholy of isolation still might not go away. I guess what I'm trying to get at is, or just ask the question about: Is there a way of making the loss *the thing* that actually is centered as that positive thing because of that?

It doesn't mean that you don't feel adrift in certain moments. It doesn't mean that you don't feel angry, you know, about not knowing.

John: I'm glad you mentioned an existential process earlier because what you just said is incredibly existentialist, in my view. Simone de Beauvoir says there is no biological nature to gender and Jean-Paul Sartre says there is no biological essence of being human. Rather, we create the essence of gender and humanity through our choices. Is something similar happening here? I just wonder if in not feeling empowered or limited by biology, through that kinship of loss, there are—like you just said—some positive opportunities. I hate to put it that way, to call loss opportunity from my privileged position, so please challenge me.

Shannon: Yes, that is the positive opportunity. I think we have to talk about the positive, because if we don't then we're always going to be a grief-averse culture, right? There is this other side. You talk to any therapist worth their salt, they'll tell you that you can't experience joy if you don't experience grief. You can't. It just doesn't work like that. You know? It's not a healthy orientation that we have in the West toward grief. That is part of the work that I'm trying to do—through theorizing this kinship of loss with transracial adoptees, who are not the only community at all that I could see this working for. That's just my community and that's my experience. It's positive because you understand that you're not alone.

At one point in my first novel, *See No Color,* the protagonist Alex—a young, mixed-Black transracial adoptee—basically says, "All my life I thought of myself as a freak." And then she meets this transracial adopted woman, Kara, and Kara's maybe ten years older than her, and Kara gives Alex context, information, and connection that Alex had been lacking. Then Alex is like: "Now I realize that maybe I'm not a freak at all. Maybe I'm just an archetype and maybe I'm just part of this outlaw group, and maybe we weren't responsible for anything that had happened to us." That's profoundly liberating to realize—to be aggrieved and for that to be acknowledged.

John: And then grieved and then understood.

Shannon: Like, "Yeah, it's not my fault." You know, it's like my Korean adoptee friends. They can't be part of Korean American identity, because Korean identity is speaking Korean. They can't speak Korean, a bunch of them. That's not their fault. If you can't speak Black English, that's not your fault growing up. And I'm like, if you don't have friends that look like you, that's probably not your fault.

Right? It's like these things that we carry around as humans, as a ball and chain.

John: And shame gets internalized and you believe everything's your fault?

Shannon: But you don't even know that you're doing it. That's the insidious part, right?

It's like, you're blaming everybody for everything and you don't even know that you're approaching everyone in your life with the fricking ball and chain. And they're like, "What is going on with you?"

John: It strikes me that much of this is *work,* albeit very liberatory and cocreative work in all of the ways you have spoken about. This is the work of kinning, right? And you have done so much kinning work in your life as a transracial adoptee person, theorist, novelist. And you are also a mother (and I'm curious about whether there's much work on the children of transracial adoptees). How is mothering part of the practice of kinning for you, since that connection with your biological children hasn't been severed for you as mother in the way that it has for you as a daughter? You have done a lot of personal kinning in your time. For example, you have tried to reconnect as tourist and then as a partner and daughter-in-law in African countries like Ghana and Liberia. But I am curious about your thoughts on the practice of kinning in terms of how that involves the next generation for you.

Shannon: Oh my goodness. OK. Just tiny, teeny, teeny, teeny-tiny questions, John [*shared self-deprecating laughter*].

John: OK, let me make this less abstract. You did a lot of kinning in Minneapolis upon finding that transracial adoptee community. Yet the way I've seen you build your family is an incredible example of the practice of kinning, too.

Shannon: Right, let's look at Black cultural practices. Like the cousins thing that's so widespread among Black folks. Friends will be just like, "Yeah, here's my cousin Elaine," and I'm just like, "I didn't know you had a cousin."

"No, she's my play cousin. You know, we grew up together." That's a thing that's recognized, you know, and it's not just Black folks. Indigenous folks, pretty much anybody nonwhite has some version of that.

In Sarah Schulman's work—she's a queer theorist and she has this book *Conflict Is Not Abuse*. In one chapter, she talks about the danger of the family unit. She says something like, "I quake, I quake in fear at the chosen family," because, for her, friends and colleagues are the center of her liberation, because of the things that she has seen nations and governments and individual people do on behalf of the family unit. That's something that stopped me. That view is not somewhere that I necessarily want to go as a mother, but I just think that's something to think about.

John: I'll just say one really cool thing about your answer there. You're still centering your view of kinship on the nonbiological, which seems very empowering. As much as you love and give to your children, even when I opened the conversation to the topic of motherhood as kinning, you still went to a nonbiological kinship answer. It is a deep expression of kinship, what you say about Schulman, and still decidedly nonbiological.

Shannon: Kinship will never be strictly biological for me. I will never be able to do that. That will never work for me, what it means in terms of my parenting. But I love your question about the children of transracial adoptees, because there's not a lot of work on that. My parents are now grandparenting Black children who are children of their adoptee. Again, there's just not a lot of work on that. There's not a lot of theoretical work on it. There's not a lot of social work on it. There's not a lot of literary work on it. And that's not good because there's more families at that stage, you know? But for me, the thing about my experience as a mixed-Black transracial adoptee that has impacted my parenting is that I'm hyperaware of power dynamics in the family and how much power I have over my children. Like all the time in our family structure, in our interactions, you know?

The narratives that we tell about our kinship leads us to just believe willy-nilly in a biological, genetic understanding of kinship. And they don't tend to see the constructedness of kinship and power in families. They don't see that. And that causes a lot of problems, like a lot of problems, because one of the only ways that you can get people to check their power voluntarily is by them acknowledging that they have it.

John: In addition to checking people with that privilege of biological kinship, are you also freeing them from biologically based kinship?

Shannon: Yes, because it's all about narratives. There's all these stories. Yeah, if I (or all of us) don't have to be defined by these Western notions of kinship, which are so located in biology, then someone else does not necessarily have to have a certain kind of relationship or interaction with their physically abusive father, or whatever it is. And I can . . . [*Shannon's children run into the room where she is on this Zoom interview to show her something they made. She looks off camera to see their creation and says to them, "That's so beautiful." Shannon turns back to the camera to finish her thought*] I can say, instead, "My father is the hill over there."

NOTES

1. Shannon Gibney, "Kinship between Transracial Adoptees: A Case for the Kinship of Loss," in *Ethnicity and Kinship in North American and European Literatures*, ed. S. Schultermandl and K. Rieser (New York: Routledge, 2021), 160.
2. Gibney.

POETS AND HUNTERS

Brian Calvert

Five thirty in the morning and my watch alarm taps me out of a mountain sleep. Deep breaths in a dark tent. Pitch black. New moon. Two hours to sunrise. Find the headlamp, switch it on, leave the warmth of the bag to squirm into my clothes. Grab the pack and the rifle, zip the tent closed, set out for the eastern ridge, which, come daybreak, will allow a wide view of the valley, where the creek runs from the high bowls, through beaver ponds and meadows, and where perhaps at dawn the elk will deign to drink before they slip back into the dark timber and lie down for the day.

By the beam of the lamp, the fog of my breath leads me down a narrow trail, through a stand of fir and aspen, into open country, sagebrush and dry grass, and after a few hundred yards, I leave the trail and head uphill. My legs feel the steep of the hill, the weight of the pack, the weight of the rifle. It was my grandfather's, then my uncle's. The stock is carved with a diamond pattern that Uncle Gary put there as a teen. It is of light caliber and heavy history.

I have not gone far when I run into a thicket of scrub oak. I had meant to skirt it but missed the way through, and in the dark under the stars, I'm not sure which way to go. The plan was to walk a full hour up the ridge. I've barely been walking twenty minutes. I trace the edge of the thicket, unwilling to enter. That world is for rabbits and bobcats, mice and birds. I settle on staying put. I rest the rifle on its bipod, pull the binoculars and a down parka from the pack. Turn off the lamp, sit on the pack. Wait.

I grew up in a small town in western Wyoming, on high plains of sagebrush, the traditional land of the Eastern Shoshone. Both my parents and all of my grandparents were born in Wyoming, Anglo settlers in an uncharitable place of wind and cold, who depended on hunting to fill the freezer. Every fall we hunted—grouse, goose, duck, deer, pronghorn, elk—ruthlessly and practically: anyone who could get a tag got a tag, and we filled those tags. As a boy, I walked the woods and hills, waded marshes, hunkered in blinds and hollows, often cold, often bored, always trying not to talk or shiver too much, watching and hearing the world.

At fourteen, I killed my first deer, a four-point buck moving through a stand of yellow aspen, a precise shot to the neck, my chest swelling and heart sinking at the death and blood of the animal. In my family, there was no philosophy, no meaning, attached to this kind of killing. This was, my grandfather would say, "God's country," and it provided means of survival and that was enough. Still, I was taught you could read the land, imagine the thoughts and ways of animals. Grandpa could catch or kill almost anything with a preternatural ability to find quarry. When I was a boy, he would take me along as he checked his beaver trap lines; taught me to set traps, too, to squeeze open the jaws of the strange, malicious metal, to scent the trap, erase all signs of our passing. There were scents and calls and other tricks, but if you couldn't read the land, they were useless. The first twenty dollars I earned was from a muskrat pelt I sold him, one I'd skinned and stretched myself, in his cool garage, the musky odor of animal oil all around, under the gaze of a stuffed golden eagle. Grandpa ran a fur business, and he live trapped minks, foxes, and bobcats, bred them and raised their young for pelts. My sister and I would bottle-feed bobcat kittens, never understanding the fur trade they were fated for.

Once I entered high school, I gave up on hunting, moved on to literature, then spent much of my adult career as a journalist abroad, writing from places of conflict and violence. When I had seen enough of that world, I settled into a job in western Colorado as a magazine editor. By then, my grandparents had passed, and my mother too, and after many years away, I felt bereft, dislocated. By then, too, the ecocide was obvious, the planet overheating, a mass extinction under way. I knew deep down that my family's "way of life," my entire upbringing, had been built from a system of conquest and consumption, of goods, animals, people, land, and labor, leading the world to calamity. I felt a deep sense of shame, but in that a responsibility, to look into my heritage and salvage something. I had grown up with a love of wild places. Surely I could find something good in them again. I retrieved the old rifle, then in the care of my father, and set out to relearn to hunt, and maybe to learn something more of what it means to exist.

I would hunt elk, a Rocky Mountain subspecies of *Cervus canadensis*, a large member of the deer family, known to European and Native peoples as wapiti. The elk is an Old World animal that found its way across the land bridge, and so North American elk share a lineage with the red deer, Irish elk, and other antlered ungulates. They thus share a poetic kinship with creatures of myth and legend: with Actaeon, whose transgression is to see the goddess Artemis, the moon maiden of the silver bow and wild mountains, naked and bathing, who, offended, transforms him into a stag and sets his own hounds upon him; with the emblem hart of Heorot Hall, where Grendel prowled and battled Beowulf; with Cernunnos, the antlered god of Celtic forests and fertility; with the white stag whose antlers held a crucifix, leading the Roman soldier Eustace to Christianity and sainthood; with the deer cults of Thrace; with the divine messengers of the Shinto; with the golden deer of the Ramayana; and with Eikthyrnir, who stands over Valhalla, browsing the highest boughs of Yggdrasil, the World Tree, and from whose antlers flow all waters.

The dark gnaws, the ground a cold hunger that steals heat from my body. Poets and shamans have sung stories ever since we had language to sing and firelight to sing by. My ancestors, after a full day of butchering carcasses and digging roots, of hiding from jackals and lions, gathered at night around flickering flames, singing stories to the stars. In the dark, I strain to hear them. A fire now would be a welcome thing. In the stillness, the cold creeps in. I wrap my arms around my knees, flex my arms, my legs, my abs, to get the blood flowing. The toothed cold nips at my fingers and ears. At last, the dark relents, and the shapes of the hills and mountainsides emerge. I am facing west, my back to the mountain, so there is no rosy-fingered dawn—just the palest of gray light seeping into the world. I close my eyes and take deep breaths and dissolve into the coming light. A small prayer for elk to appear: *please show yourselves to me*. Silence. Cold. Then, there, at the edge of the aspen, near a spring, a cow elk. Another. And another. They drift out of the trees, ghosts in the twilight. Like all living things, elk exist in a weave of patterns. The bulls shed and grow antlers in an annual cycle of sunlight and season. In the fall, in the rut, the bulls battle for the right to a harem, bugling and challenging one another to antler-locking combat, while the cows, matriarchal in nature, stay in small groups that will gather into larger herds in the winter. All summer, the bulls stay in the high country, eating foods rich in minerals to rapidly grow massive racks, long symbols for virility. The cows find their way to nutrient-rich plants that help them nurse newborns. By fall, all elk, and their hunters, must balance their needs for the coming winter against the high levels of activity that must take place to perpetuate the species.

This morning, we are past the September rut, and although it is cold, there is not yet snow. The elk are in their daily cycle, a ritual, moving at dawn to find water and grazing before drifting back into the dark timber to bed down for the day. A hunter must understand this pattern and combine it with more pattern—minute patterns like fresh scat and tracks, antler rubs on trees, forage, and water, with larger patterns of landscape and seasons, the forest edges and meadows, the weather, the phase of the moon, and cloud cover—and place themselves in the midst of these natural patterns in hopes of aligning with them. When I am hunting, I am reading patterns, synthesizing information, and flowing through the natural world in a kind of trance. The eyes see wider, and more, the ears stay perked, the nose becomes more useful. The reward is the appearance of the elk, which lights up the mind with bright pleasure, the same pleasure that would have enlightened our fruit-eating primate ancestors, the same pleasure that lights up our artistic sensibilities. The right poetic line read at the right time ignites delight in some same part of the mind. Leaf, leaf, flower, fruit. Tree, tree, glimmer, elk.

The work now is to stalk the elk that have shown themselves to me, to move within rifle range. In the dim light, I skip off the hill, knowing I am perhaps half a mile distant from the herd, who have settled into a grassy hollow where a spring runs through the aspen. I measure the landscape: a creek between the elk and me; and, on the other side, rolling hills, pocketed with a copse of fir and a sweeping stand of aspen. I can move from one pocket of trees to another. I carefully cross the creek, wary of the noise of rushing water, the song of the stream whirling from the stones of the bank. I skirt behind the first small stand of fir, then peer out from the edge. Three elk there, five hundred yards away—a small bull, a cow, a yearling. My license is for a cow, so I focus on her. She is grazing on bunchgrass. I imagine myself in her world, not just through her eyes, but her mind. What can she smell? Hear? Taste? What is the wind doing? Here I become elk.

This is our magic, we humans, we without fur or fang, the power to become Other, to copy nature—in birdcalls and bugles, in song, poetry, and painting. All art is born from this wonder. Mimesis. I think hunters were the first artists; shamans, the first poets. The poetry of Orpheus, the tamer of beasts, could captivate not only men but also animals, even gods and ghosts. The poet's mother was a Muse, the embodiment of poetic thought, Robert Grave's White Goddess of birth, love, death—nature—the "single poetic theme." Painted on the wall of a cavern called the Sanctuary, at the Cave of the Trois-Frères, in Ariège, France, is the thirteen-thousand-year-old figure of, if we can believe the sketches, a "sorcerer," an antlered humanoid, a first god, a sign of hunter as artist. This mimetic function, sadly mostly relegated to art in Western culture, appears across other cultures and myth.

I lived many years in Cambodia, where animism melds with Hindu myth and Buddhist thought, where tree, rock, and water spirits share the world with ex-soldiers, rice farmers, monks, and hungry ghosts. This is common, and beautiful. In his observations of the Yukaghirs, hunters of Russia's taiga, Rane Willerslev notes that "their entire cosmos is in effect a hall of mirrors, as various dimensions of reality are conceived as replicas or reflections of others." Mimesis. "When the hunter seeks to bring [a moose] into the open by mimicking its bodily movements, he is inevitably put into a paradoxical situation of mutual mimicry," Willerslev writes. "As a result, the bodies of the two blend to a point that makes them of the same kind . . . [so that] hunters are both human and the animals they hunt, both predators and prey."[1] This comes as no surprise to me. I believe it is one of the reasons that I, a poet, hunt, for I believe the duty of both is to dissolve into the worlds of other beings.

From my vantage at the edge of the firs, I measure the distance to a small hillock, where an old tree has fallen over, its roots a silent-scream gnarl of gray wood and granite stones. If I can reach that place, I will be within comfortable range of the elk. I move carefully, dipping into a crouch, then a crawl. The summer rains put some high grass here, which can help confuse the elk. As I move, I watch. The bull senses something amiss, raises his head, looks in my direction, his antlers spreading behind him like a sculpture or princely crown. I still myself, wait several heartbeats. The bull waits, too, then goes back to grazing, and I move the last stretch to the fallen tree. I unsling the rifle and lie down, carefully releasing each leg of the bipod. I tuck the butt of the rifle into my shoulder, feel the earth push against my elbows, cradling me, absorbing me. My heart and chest push into the earth, and the elk keep grazing.

I lean into the rifle and set my right eye against the scope. I find the elk in the dreamlike world of glass and optics, larger now, somehow unreal. I level the dot that is the sight on the cow. From here, perhaps 250 yards away, the best idea is to aim for the heart and lungs. I have never wounded an animal, a prospect I forever dread on a hunt. To wound an animal is the cruel breaking of an ancient compact. Everything now is in the breathing. Breathing and being. I have spent the morning as the elk, am now soul-bound to this elk through the scope and the lethal potential of the rifle. She walks a few steps in my direction, turns a quarter away, then puts her head down to the ground, her nose snuffling the grass, her lips, mouth, and teeth tasting the grass, nibbling the grass from its roots, chewing the grass in the autumn cool. The creek is a whisper somewhere behind me, and an early morning breeze crosses the meadow, ruffling leaves and brushing me and the grass and the elk without favor. In this still moment, we are all here in the in-between, all one thing, akin on the precipice of beauty and violence.

NOTES

1. Rane Willerslev, *Soul Hunters: Hunting, Animism, and Personhood among the Siberian Yukaghirs* (Berkeley: University of California Press, 2007), 11.

DISPATCH FROM THIS SUMMER: *LYMANTRIA DISPAR DISPAR*

Elizabeth Bradfield

Frayed, moth-eaten, vulnerable. Those Florida dancers
gunned down & my young self coming out dancing & pathetic
fallacy (*dispar dispar*) crawls all over June's fresh oaks,

gnawing them to a February canopy. The news, bad
oracle, gnaws fact & rumor. Above, unrelenting
mastication, defoliation. *Lymantria*, "destroyer," all else gone,

you hump up even the stiff needles of pines. What will happen
come winter, no sun stored? Should we spray? Should
we shun social media? Avoid large aggregations? How

hot the birds must be, so now unshaded in their nests. (Guilty
thrill of peering down on them, black-billed cuckoos calling.)
Other wings. The white towering stagecraft of angels

sentry at Orlando's mourning. We consider what it would take
to pick the trees clean. Could we? The bark the grass the ground
writhes. In a grove in China, a grim documentary:

honeybees gone, people pollinate fruit trees by hand. I
twitch away
from one caterpillar dangling from its thread, hanging by
the silk that brought it here, to the New World, to Massachusetts

even, because some merchant in 1869—while Grant
took the presidency and Elizabeth Cady Stanton spoke
before Congress and the Golden Spike was hammered

into Utah and the South fumbled through what's called
Reconstruction—thought *crop, harvest, riches* & hoped
the long, expensive trek to mulberry unnecessary. We gnaw

through news feeds. We post & share, unsure
if we are offering or consuming. In the forest, a constant
heavy frass. On my side of the river, healthy trees. Oak leaves

thick and dark. In the dance clubs near me, there is
dancing. But introduction, dispersal. In the week
after Pulse, in Massachusetts, 450% more guns like that gun

were sold. If you can stand to walk a narrow path through
the leafless
forest, you can arrive at a circle of water that will allow your body
to be beautifully held, whoever you are. It's true, you'll have
to return

by the same path, go back through those apocalyptic trees. If
I had waited a month to begin this poem, I would have begun
with the releafing, fuzzed red growth in late July. Is it too late?

Now, plastered to bark, the russet humps of eggs that I scrape
with a stick
—vengeful, hopeful, despairing—even as they are being laid.

—North Truro, 2016

HUMANS' NEXT OF KIN: *BLACK PANTHER*, OUR INNER GORILLA, *EARTHRISE*, AND EARTHLING NECESSITY

Kimberly Ruffin

When I started working on the concepts surrounding human animality, the list of names included Atatiana Jefferson, Trayvon Martin, Sandra Bland, Eric Garner, Michael Brown, and Botham Jean. By the time I began writing the essay you're reading now, there were more names to add to the list: Breonna Taylor, George Floyd, Ahmaud Arbery, and Elijah McClain. While the response to the list of Black folks dead due to white supremacy feels different now, the movement to recognize Black people as "humans" still suffers from the brute force of white supremacy.

So as a Black woman scholar who has chosen "our inner gorilla" as one of the subjects of this essay, I don't take up this subject or come to my conclusions lightly. To those whose animality is perceived as superior, racism gives a license to kill. I, my family, and the larger community of people of African descent pay that cost with our bodies, whether it's a macroaggressive murder or the accumulated physiological corrosion of daily microaggressions. The cost results in a racial mortality gap, in which "Black men have the lowest life expectancy and Black women are 3 to 4 times more likely to die during pregnancy than White women."[1] As the antiracism scholar Ibram Kendi puts it, "The Greatest White Privilege is Life Itself."

Yet here I am writing on the topics of humans, gorillas, and earthlings in light of the fact that, in the words of Philip Butler,

"Black folks in the United States are just beginning to explore what it means to be both Black and human within the bounds of 'equality.'"[2] Black protesters in the 1968 Memphis Sanitation Workers' Strike evoked the language of humanity by wearing placards emblazoned with the simple but definitive statement "I AM A MAN" (fig. 1). Donning human uniforms (i.e., suits) meant to command respect, these protesters confronted directly one of the founding lies of white supremacy: Black "people" are something less-than-human within a context in which whiteness is misunderstood as the pinnacle of "human" existence.

Figure 1. Civil rights activists are blocked by National Guardsmen brandishing bayonets while trying to stage a protest on Beale Street in Memphis, Tennessee. The marching demonstrators, who are wearing signs that say "I Am a Man," are also flanked by tanks. Photo credit: Bettmann. With permission of Getty Images.

Lest we think this is a particularly American or expired problem, international soccer matches provide a relevant example of the global consequences of being marginalized from the human community and "relegated" to subhumanhood. Throngs of

European fans are willing to hurl racist monkey chants and bananas, all in an effort to debilitate elite African-descended athletes. These twenty-first-century racist acts were so popular that Russia even appointed an "anti-racism inspector" for the 2018 World Cup. The mostly African-descended, French team that took home the 2018 World Cup trophy had the last laugh on racism. Bananas and racist jeers certainly did not stop them from reaching the zenith of sports achievement.

But they have stopped other Black footballers in their tracks. Notably, in 2017, the Brazilian football player Everton Luiz, who was playing with his Serbian team members, left a soccer game in tears. He later remarked: "I took 90 minutes of racist abuse and other insults from the terraces and thereafter I found myself in a cauldron of emotionless individuals who charged at me when they should have protected me. I want to forget this, refocus on football and urge everyone to say 'No' to racism."[3] What is the aftercare for a situation of racial trauma? Should his team and their coaches get sensitivity training? Do you ban or fine the rival fans? Does Everton Luiz go into intensive talk therapy?

More broadly, if one does survive racially traumatic events, is it possible to remove the impact of those events from one's mind, body, and soul? How do these experiences reverberate for African-descended people, spilling out and complicating multiple areas of their lives?

My most vivid experience with this phenomenon came when I was coteaching a glorious twenty-eight-day experiential learning trip in the beautiful city of New Orleans. Waiting for a bus, a student who is also Black and I were stupefied when a blond-haired white man in a convertible whizzed by us and yelled, "Can I get a banana?" What's the experiential learning lesson there? This was sixteen years ago, and yet this five-second encounter with a perfect idiot still lives with me. My mind's eye can put me there in an instant.

Maybe it is clear that ending the racially stigmatized association between nonhuman primates and people of African descent will at least enhance Black people's ability to sustain their mental health, but what relationship does this have to environmental sustainability and social equity? The motion picture *Black Panther* has helped me realize that while fighting for "humanity" is essential for Black health in the current public sphere, it is only a start. Because of its embeddedness in white supremacist thinking, relying on the term *human* to achieve social justice does more harm than good. We can advance beyond the dangers of human-centric thinking by upholding our inner gorilla and exploring humans' next of kin: earthling. *Black Panther,* which elevated gorilla (and thus, nonhuman) imagery, led me to realize that earthling is a necessary identity we all must embrace to ensure our place on this planet.

Discussions of topics as raw as racial injustice and environmental despair can be dispiriting in the absence of a complimentary vision of possibility. The motion picture *Black Panther* gave me an aesthetic experience filled with possibility, opportunity, action, and empowerment, particularly through the character of M'Baku, masterfully played by actor Winston Duke. Duke's breakout performance as M'Baku, and the powerful sight and sound of the actors who played the Jabari warriors, yielded the possibility and opportunity to envision a group of African-descended people associating themselves with a gorilla without white supremacist stigmatization. In the movie, the Jabari tribe worships a gorilla god, calling it Hanuman (which is also the name of a Hindu god associated with a monkey), and their worship is unfettered by the objectifying lens of racism because it takes place in the fictional, and uncolonized, world of Wakanda. They revered their inner gorilla and, because of this reverence, played a crucial role in the narrative.

This provocative imagery did not go unnoticed. In fact, empowerment born of this animal association in *Black Panther* prompted the viral sensation of the "M'Baku Challenge." Untold numbers of fans recorded themselves saying M'Baku's lines from the film and shared them on social media. Everyone from Black youngsters using towel capes and pool noodles to Black adults recording themselves in cars re-created the powerful utterances of M'Baku and the Jabari tribesmen. There were so many responses to the challenge on social media that "best of" compilation videos emerged of African-descended people reveling in the undeniable artistic energy of the characters. It is glorious to see Black people inhabiting an aesthetic vocabulary where an association with an animal as mighty and commanding as a gorilla is a point of pride, play, and power.

However, when I first watched *Black Panther* and saw the Jabari men exit the cave with their thunderous chants, I was confused. Along with millions of others around the country, I was excited about the film's premier. To ensure that I could get the best seats available, I arrived early to the movie theater with expectations that the film would live up to all the positive publicity. But hearing the vocalizations of the Jabari men (unmistakably mirroring nonhuman primates) and seeing their "primitive" animal skin costumes, I was furious. *Not this crap again!* I thought. As one of the most thoroughly hyped Black films of my lifetime, I thought: *Surely* Black Panther *wouldn't reproduce this negative animal association that continues to haunt Black people.*

As I watched the movie, I hadn't yet realized the extent to which the screenplay's cowriters had adapted M'Baku from the original character who appeared in the Marvel Comics. In the comic books, M'Baku's supervillain identity was "Man-Ape." This was the image that I feared: a physically domineering, primitive, murderous, supervillain with superhuman powers that could be directed for good only when associated with someone more "human."

How do you get "Man-Ape" out of the racist quagmire of his original creation and into his Afrofuturistic re-creation in Black Panther? This is a question best answered by the cowriters, Ryan Coogler and Joe Robert Cole. By the time they had finished writing and Winston Duke had finished acting, M'Baku had been transformed.

Sometimes art provides the aftercare from racial trauma you did not know you needed. Sometimes aesthetics deliver something you could not anticipate. Sometimes it is a paradigm shift.

Cole explains: "On a fundamental level, one of the amazing things about Wakanda is that it's an African nation that is wholly self-determinant. It's never been conquered. It's never been colonized or overrun. It decides what it is. We're at a time when people of color are asserting and affirming their own self-determination. I think that disruption of the kind of prevailing paradigm that we've been living under is really resonating with our movie."[4] *Black Panther* is chock-full of paradigm-busting images. The film's deployment of M'Baku as a transformed "Man-Ape" signals not only the busted paradigm of racial inferiority within the category of human but also the limitations of the category of "human" itself. In doing so, *Black Panther* gave me an opportunity to think carefully about that monkey that's been on the backs of Black people for so long.

The mountain-dwelling, gorilla-associated Jabari tribe simultaneously washes away the debris of white supremacist human hierarchies and brings into question the presumed superiority of the human being in species hierarchies. The Jabari people are the critical outliers in Wakanda who question the technophilic orientation of the rest of the Wakandan world. Commenting on the scientific leadership of Shuri, T'Challa's younger sister, M'Baku exclaims, "We have watched with disgust as your technological advancements have been overseen by a child who scoffs at tradition." The technocentric parts of the Wakandan universe mirror contemporary society's preoccupation with technological answers to environmental problems. These answers rely on presumed human ingenuity rather than cultural changes that reinscribe our planetary dependence. A white supremacist fixation on "humanness" and the idea of human species "exceptionalism" detaches *all of us* from our life-support system. M'Baku gave me a clear vision of a people reveling in animality in ways that love human beings back down to planet Earth, in all its glory.

My pre–*Black Panther* experience of the association of nonhuman primates and Black people polluted my eco-imagination with shame, kept me from an affinity with other primates, and wasted my time with the whole question of how "special" and "different" human beings are from other animals.

In *The Origin of Others*, Toni Morrison asks: "What is the nature of Othering's comfort, its allure, its power? Is it the thrill of belonging—which implies being part of something bigger than one's solo self, and therefore stronger? My initial view leans toward the social/psychological need for a 'stranger,' an Other in order to define the estranged self."[5] I contend that racial hierarchies function in the same way that species hierarchies do: They estrange people from themselves and give people a false sense that they are above and separate from their fellow earthly kin. In doing so, focusing on human exceptionalism encourages us to distance ourselves from crucial knowledge: we, like other animals, are earthlings. Human exceptionalism supports the mind-boggling expectation that, even if we cannot come up with the technology that allows us to stay on Earth, we can pick up, use our technology, and go somewhere else.

Those working for sustainability and social equity have to strive both for a world that sees the full humanity of nonwhite people and for a world in which humans invest deeply in their animality and their kinship with other earthlings. Black people, in particular, must be free to live with their human rights, respected in our social ecology, to fight for our right to connect to the rest of the Earth-born community, and to work to keep this planet viable as a home for ourselves and other earthlings.

Another film that reinforces my ideas about the necessary identity of "earthling" is *Earthrise*, a short film by Emmanuel Vaughn-Lee. Telling its story through archival imagery and footage and the words of the three Apollo 8 astronauts, the film chronicles the influence of a photo taken during the mission. *Earthrise*, a portrait of Earth, is one of the most reproduced images of all time.

Like the paradigm-busting *Black Panther, Earthrise* included some surprising insights from the astronauts who made this historic trip.

In archival footage, Frank Borman, who felt that the emotional weight of what they saw was best described by poets, quoted a poem by Archibald McLeish to convey that the perspective on Earth from lunar orbit is "to see ourselves as riders on the Earth together, brothers on that bright loveliness in the eternal cold, brothers who know now that they are truly brothers." Although their "escape from Earth" is bound up with Cold War politics and America's history of cultural imperialism, the astronauts expressed disappointment that their work, and the *Earthrise* photo in particular, had not promoted more of the global perspective captured in the photo. William Anders remarks that he didn't think that the Apollo program "has yet brought as worldly a view, an interlocking view to humankind, as I'd hoped." And James Lovell turns the spacecraft image on its head, saying, "We did something that ended up showing the Earth and its people exactly how we existed . . . that we were really here on Earth, a spacecraft, and we were all astronauts, whether we liked it or not."

While I am careful not to infer the astronauts' attitudes about space exploration, I also perceive that the depths of their reverence for Earth seems forever enhanced by the Apollo 8 mission. Above and beyond any categorizations imposed on us by other human beings, earthling identity thrives when we all can live out what it means to be, in Borman's words, "citizens of Earth."

Freed from some false, stigmatized identification of my Black body as a "missing link" or my Black body as some degraded version of a superior, white human body, I revel in an opportunity to express my biophilia, my love for the other living things that make up this magical, wonderful spaceship. I am free now to take the focus off demonstrating humanness in my life. I know who I am. I also know that the oppressive and exceptionalist notions of human are not enough. Now, I want to be a better earthling.

One exciting result of becoming a better earthling might be a legacy of social equity and sustainability. Alison Deming, in her book *Zoologies,* explores the emotional terrain of recognizing our full embeddedness in this planet and not letting our weddedness to a "human" identity curb our actions. Deming paints a future in which we have learned how to become good ancestors. As she writes: "Ten thousand years from now, I want someone to say of us, 'What amazing courage they had, and what spirit. How smart they were, how inventive—and how profoundly they must have loved Earth.'"[6] Loving Earth and living like we are kin with its creatures is central to an earthling identity.

Figure 2. Appolinaire and Pikin moving from the vet clinic to the new sanctuary space. Photograph credit: Jo-Anne McArthur / We Animals (2009).

With the perspective of an earthling, I am seeing things anew, especially images I once thought only promoted racist notions of Black people. There are two photos with four earthlings that I can finally delight in fully, without the menace of racism or de-Earthed humanness. The first image is one that has appeared in a number of publications: the famous primatologist Jane Goodall, with a fellow primate, a chimpanzee, looking at each other, mirroring a vocalization. As Goodall is a woman of European descent, no one has probably ever thought her "race" to be a cause to question her humanity. In the other photo, taken by Jo-Anne McArthur, who won the 2017 Wildlife Photographer of the Year People's Choice Award, the rescued gorilla Pikin and her African-descended caretaker Appolinaire Ndohoudou are shown hugging as they ride in a car (fig. 2). The pleasant expressions on both their faces convey an affinity between two beings who share a planet. Before *Black Panther*, I would have questioned the intentions of the people who voted for the second photo. In my post–*Black Panther*, post-M'Baku consciousness, in both photos I finally see four earthlings enjoying their time together, living together on an Earth they cherish.

Covering the conceptual terrain of human, animal/gorilla/primate, and earthling is crucial as we work for a world in which we can coexist, without destroying our life-support system. However, the necessity of earthling kinship calls to us revise oppressive "human" notions of ourselves and others as we search for the changes that may indeed help us promote the eco-social health of this spacecraft on which we all depend.

NOTES

1. Ibram X. Kendi, "The Greatest White Privilege Is Life Itself," *The Atlantic*, October 24, 2019, https://www.theatlantic.com/ideas/archive/2019/10/too-short-lives-black-men/600628/.
2. Philip Butler, "Making Enhancement Equitable: A Racial Analysis of the Term 'Human Animal' and the Inclusion of Black Bodies in Human Enhancement," *Journal of Posthuman Studies* 1, no. 2 (2018): 106–21, at 106.
3. See "Pro Soccer Player Leaves Field in Tears after Racist Chants," CBSNews.com, February 20, 2017, https://www.cbsnews.com/news/brazilian-soccer-player-everton-luiz-in-tears-after-racist-chants-in-serbia/.
4. Quoted in Marc Bernadin, "'Black Panther' Writers on Wakanda's Unique Celebration of Black Glory," *Variety*, February 5, 2018, https://variety.com/2018/film/features/black-panther-joe-robert-cole-evan-narcisse-wakanda-1202686413/#article-comments.
5. Toni Morrison, *The Origin of Others* (Cambridge, MA: Harvard University Press, 2017), 15–16.
6. Alison Hawthorne Deming, *Zoologies: On Animals and the Human Spirit* (Minneapolis: Milkweed Editions, 2014), 239.

ACADEMICS ARE KIN, TOO: TRANSFORMATIVE CONVERSATIONS IN THE ANIMATE WORLD

Graham Harvey

I have learnt much from generous Indigenous hosts about what it can mean to be animistic. Not all of them use, like, or accept the term *animist*. Then again, I'm a scholar of religion, and most of the people who interest me don't think they have or do religion. That's another story. The story I'm telling now is about how my academic research has been propelled and enriched not only by conversations with humans who seem to me to be animists but also by conversations with hedgehogs, mushrooms, ravens, kumara, salmon, and elbow bacteria. I confess that "conversations" suggests too much: sometimes I've talked more than I've listened . . . but sometimes I've heard more than I expected. However, one alternative term, *encounters*, is too dull and too freighted with obfuscations of the colonial invasion of the larger world.

While I've been told what animism is about by some Indigenous hosts, some Pagans, eco-activists, children, researchers, and writers, like them, I've come to understand things best by joining in ceremonies and other activities.[1] Animism, like much of life, is better done than thought or said. While this is true, the crucial thing is that animist ceremonies and lives are not solely human affairs. While learning by joining in with/in the larger-than-human community, I and other researchers have become relations—one manifestation of which is attempting to be better guests.[2] However, because enmity is also a relation, even the most disdainful critic also becomes a relational participant in the ongoing evolution of multispecies communities.

To avoid confusion, *animism* is a term now being strategically reclaimed and reused (against other derogatory uses) to engage with ways of treating the world as full of persons, human and otherwise, all more or less close kin, all deserving respect. In Indigenous contexts "respect" is not a feeling but an effort. It is helpfully glossed by Mary Black as "careful and constructive" engagement and points to inculcated and practiced attention to specific persons and to locally appropriate etiquettes of relationship. In Tim Ingold's presentation, "We are dealing here not with a way of believing *about* the world but with a condition of being *in* it. This could be described as a condition of being alive to the world, characterised by a heightened sensitivity and responsiveness, in perception and action, to an environment that is always in flux."[3]

Or as Isabelle Stengers concludes:

> Reclaiming animism does not mean, then, that we have ever been animist. Nobody has ever been animist because one is never animist "in general," only in terms of assemblages that generate metamorphic transformation in our capacity to affect and be affected—and also to feel, think, and imagine. Animism may, however, be a name for reclaiming these assemblages, since it lures us into feeling that their efficacy is not ours to claim. Against the insistent poisoned passion of dismembering and demystifying, it affirms that which they all require in order not to enslave us: that we are not alone in the world.[4]

In short, we (academics as much as anyone else) are not alone in the world, and seeking to heighten sensitivity and responsiveness in perception and action seem admirable ambitions for researchers. The fact that some academics ally themselves with the human separatist movement named "Modernity" highlights the value of re-placing ourselves in communities of multispecies inter- and intra-action.[5]

My discovery of not being alone in the (animate) world—including while conducting academic research—involves a serendipitous and unexpected journey among different human communities. Although my PhD centered on ancient Jewish identity politics, since the mid-1970s I also had been participating in the Stonehenge People's Free Festival, including some years of activism to gain free access to the stones for all. This involved fieldwork among Pagans (practitioners of new versions of Earth-respecting religions), especially the Druids (a name derived from ancient British cultures and now referring to a style of Paganism). In doing so, I joined in ceremonies and noticed that birds, animals, plants, or breezes often appeared to participate at key moments. This seemed familiar and expected but still enchanting to some of the Pagans involved. I had met animists and become intrigued.

As my research continued, I occasionally used human encounters with hedgehogs as an analogy to aid understanding of the relationships some Pagans claim to have with "the otherworld" and with "the natural world." According to traditional folktales—taken seriously by some Pagans—faeries or elves are dangerous to encounter. They visit from "the otherworld" not to grace us with romantic beauty or natural wisdom but to take away human babies and poets, among others. (W. B. Yeats's poem "The Stolen Child" illustrates this theme.) They do not necessarily do so to be nasty; they do it almost casually or even imagine they benefit their abductees. In a similar way, when we drive vehicles to celebrate seasonal festivals we are likely to endanger hedgehogs. While this might help us understand the ambivalence of folkloric knowledges, it can also encourage us to pay more attention to the unintended consequences of living ordinary human lives. I'm not aware of seeing hedgehogs often before, but since first offering this analogy, I have frequently been visited by them while preparing for or participating in ceremonies. "Coincidence" does not seem a useful explanation for something that has become familiar, expected, and still enchanting. Speaking about hedgehogs seemed to beckon them into my life. I had met animist hedgehogs and become enchanted.

In 1996, I presented at a conference on health, organized by Memorial University of Newfoundland and hosted on the Mi'kmaq First Nation reserve at Miawpukek on the Conne River, Newfoundland. The conference was followed by the community's first noncompetitive powwow. During the final honor song, as the elders and veterans danced or processed around the central drum arbor, the largest of the local eagles flew across the river and circled once over the drum group before returning to its eyrie. Locals and visitors alike exclaimed in surprise and delight. Spontaneously, people told me that the eagle's flight celebrated the local community's efforts to return to traditional ways. They took this visitation to convey something like the following: "We eagles, in the company of bears, salmon, and others, have kept the ceremonies going. We're glad you are joining in again." I had seen animist eagles doing ceremony and became excited.

In seeking to better understand Mi'kmaq and other Indigenous kinships with the larger-than-human community, I was introduced to the work of Irving Hallowell, especially his "Ojibwa Ontology, Behavior, and World View."[6] Following this lead connected me with a vibrant and expanding field of research and debate that soon became known as the "new animism" (in contrast with the interpretations and approaches of previous scholars, such as Edward Tylor in the nineteenth century, who viewed animism as a primitive form of religious understanding). I had met animism scholarship and became involved.

The new animism resonates with "turns" to things and ontology in multiple disciplines, such as the "new materialism."[7] Much of this ferment entailed dialogue with people who insisted on the thorough relationality of the world. They challenged researchers to reconsider the obsession with aloof objectivity and to take participation in the world seriously enough to attend to larger-than-human presences. As Ingold puts it:

> Science as it stands rests upon an impossible foundation, for in order to turn the world into an *object* of concern, it has to place itself above and beyond the very world it claims to understand. . . . If science is to be a coherent knowledge practice, it must be rebuilt on the foundation of openness rather than closure, engagement rather than detachment. And this means regaining the sense of astonishment that is so conspicuous by its absence from contemporary scientific work. Knowing must be reconnected with being, epistemology with ontology, thought with life. Thus has our rethinking of indigenous animism led us to propose the re-animation of our own, so-called "western" tradition of thought.[8]

The contexts in which I have been encouraged to "rethink indigenous animism" and to reanimate the traditions of my discipline (for example, by attending to other-than-human actors) have entailed the privilege of spending time with animistic Anishinaabe, Māori, Mi'kmaq, Sámi, Yoruba, and other Indigenous hosts. I have continued to "hang out" with Pagans, and I have met some animistic Baha'is, Buddhists, Christians, Jews, Muslims, and other religionists.[9] But in most of these contexts, it is conversations across species boundaries that have most expanded my appreciation of what animism involves. Almost every time I have talked with animists, there has been at least a moment in which people have engaged with some other-than-human person or kin. They might casually touch a tree in passing; nod greetings to a bird; offer gifts of tobacco, sage, or kinnikinnick to venerable rocks; or pause our conversation to silently or gesturally acknowledge our presence in the home of other species.[10] In more ceremonial contexts, it is rare for other-than-human kin not to be invited or acknowledged as participants. Sometimes, indeed, they are central to the conduct of rites, illuminating Gary Snyder's inspired incantation, "Performance is currency in the deep world's gift economy."[11]

Ingold's encouragement to reconnect "thought with life" arises from the regular insistence of Indigenous people that knowledge brings obligations and requires engagement. Conversations among animists has, therefore, necessarily involved adjustments of my living in the world. Initiatory experiences in the company of (magic) mushrooms and colleagues in "the university of the hedge" have, for instance, demonstrated the prescience of William Blake's assertion that "if the doors of perception were cleansed every thing would appear to man as it is, Infinite. For man has closed himself up, till he sees all things thro' narrow chinks of his cavern."[12]

The cleansing purification offered by mushrooms and other powerful plants is as vital as the altered perception that has been emphasized most by enthusiasts of Blake's provocation. Whether or not mushrooms reveal "the truth," they certainly clear away the fantasy of human exceptionalism and open up vistas of the captivating "shimmering" of interspecies relations.[13] The flight of ravens, pulsing colors of yew bark, patterning of leaves and stars, and the riffing of multispecies musics—all invite commitment to the unfolding potential of lives threatened by modernity's assault. There is no place here for the scholarly distance that spearheads the fantasy of human difference.

There are also more everyday relations and lessons. From the late Māori scholar Te Pakaka Tawhai, I have learnt to attend to the kin we eat. Tawhai stated that "the purpose of religious activity among my people is doing violence with impunity."[14] He illustrates this by noting that such "religious activities" arise from the recognition that sheltering and feeding guests (a cultural requirement) requires the cutting down of trees and the digging up of kumara (sweet potatoes). While these are cosmically and culturally intimate kin, their transformation into meeting houses and food allow everyday life to emerge and diversify. So, when trees become meeting-house pillars, they continue to make space in which negotiations and interventions can take place. Without these interventions, Earth and Sky, floor and roof, would be veneered together, permitting no unfurling of lively potential. When kumara migrated to Aotearoa/New Zealand with Māori, new gardening practices evolved to enable the new arrivals to survive in the novel conditions of a cooler environment. Thus, kumara are among the ancestral life givers and taking their life is an act of intimate violence that requires religious activities. Knowing this entails the obligation to reflect on everyday relations with food persons (in farming, gardening, shopping, eating, and composting). This may result in the expression of gratitude and seeking of permission to take and consume lives. We can theorize relationality at more or less intimate or massive scales, but it is in eating those with whom we coevolve that we meet our closest kin.[15]

Maybe we have yet closer kin. Our relations with myriad bacteria make us symbionts or holobionts.[16] Bacterial DNA exceeds human DNA on and within "our" bodies.[17] They collaborate with our mothers and other relations to aid our growth before and after birth. They make nutrients available from the food that we (and they) consume. They would aid our dissolution into nutrients useful for other life, if we were not so obsessed with individualized, sterile, and separatist deaths. Our relationships with bacteria provide a perfect example of what Lynn Margulis identified as "symbiogenesis," the standard process by which evolution forms everything relationally.[18] Among these minute kin, my favorites are the ones whose preferred dwelling place is our elbow crooks. Apart from whatever they do in such places, they have enabled me to challenge the escapist notion that we need remote places in which to engage with other species.

As previously noted, gaining knowledge also entails obligations. During fieldwork with the Reassembling Democracy project (funded by the Norwegian Research Council), I met a Sámi man by a river surrounding the Riddu Riđđu festival site. We had a casual but serious conversation about the river, which seemed likely to flood the surrounding land. When I expressed concern for the festival, he explained not only how this was a result of climate change (mountain snow melting faster than ever before) but also that the rapid flow and cold temperature of the water would prevent salmon and trout entering the river from the fjord. If they did not manage that soon, they would fail to breed and thus disappear from the river forever. Whether or not his analysis was correct, it was his concern for the well-being of the fish that made a lasting impression on me. He might also have worried about human subsistence and Coastal Sámi culture, but what he talked about was the shame of what human consumerism has done to other-than-human kin such as the fish and the waters. I have written elsewhere about Māori efforts to liberate waters flowing from springs on Mount Ruapehu, which offer similar lessons about Indigenous kinship within the larger-than-human community.[19] Understanding the conditions by which waters flow and fish breed obligates us to act in concert against rising threats to all lives.

Just as animism is about transformations caused by specific acts of relating with kin (human or otherwise), so academia evolves through interactions among all our relations. The delusion that scholarship can be a detached activity suggests the triviality of "knowledge" that need not obligate us to act. In reality, scholars are kin, and there are urgent demands on us that require animated engagement. I have learnt among animist kin (hedgehogs, mushrooms, ravens, kumara, salmon, and elbow bacteria) and become obligated.

NOTES

1. For examples of this, especially in terms of the impact of writing, see David Abram, *The Spell of the Sensuous: Perception and Language in a More-than-human World* (New York: Vintage, 1996).
2. Graham Harvey, "Guesthood as Ethical Decolonising Research Method," *Numen* 50, no. 2 (2003): 125–46.
3. Mary B. Black "Ojibwa Power Belief System," in *The Anthropology of Power*, ed. Raymond D. Fogelson and Richard N. Adams (New York: Academic, 1977), 10.
4. Isabelle Stengers, "Reclaiming Animism," *e-flux* 36 (2012), 9, http://worker01.e-flux.com/pdf/article_8955850.pdf.
5. Karen Barad, *Meeting the Universe Halfway: Quantum Physics and the Entanglement of Matter and Meaning* (Durham, NC: Duke University Press, 2007).
6. A. Irving Hallowell, "Ojibwa Ontology, Behavior, and World View," in *Culture in History*, ed. Stanley

Diamond (New York: Columbia University Press, 1960): 19–52.

7. See, for examples, Rosi Braidotti, *Metamorphoses: Towards a Materialist Theory of Becoming* (Cambridge, UK: Polity Press, 2002); Karen Barad, *Meeting the Universe Halfway*.

8. Tim Ingold, "Rethinking the Animate, Re-animating Thought," *Ethnos* 71, no. 1 (2006): 9–20, at 19.

9. Behavior encouraged by Clifford Geertz in "Deep Hanging Out," *New York Review of Books*, October 22, 1998, http://www.nybooks.com/articles/archives/1998/oct/22/deep-hanging-out/.

10. See Lawrence W. Gross, *Anishinaabe Ways of Knowing and Being* (New York: Routledge, 2014).

11. Gary Snyder, *The Practice of the Wild* (New York: North Point Press, 1990), 75. See also Ronald Grimes, "Performance Is Currency in the Deep World's Gift Economy," in *Handbook of Contemporary Animism*, ed. Graham Harvey (New York: Routledge, 2013), 501–12.

12. Andy Letcher, "Psychedelics, Animism and Spirituality," in *Handbook of Contemporary Animism*, ed. Graham Harvey (New York: Routledge), 341–57; William Blake, *The Marriage of Heaven and Hell* (1793), in *Blake: Poems and Letters*, selected by Jacob Bronowski (Harmondsworth, UK: Penguin, 1958), 93–100.

13. See Deborah B. Rose, "Shimmer: When All You Love Is Being Trashed," in *Arts of Living on a Damaged Planet*, ed. Anna Tsing, Heather Swanson, Elaine Gan, and Nils Bubant (Minneapolis: University of Minnesota Press, 2017), G51–G63.

14. Te Pakaka Tawhai, "Maori Religion," in *The Study of Religion, Traditional and New Religion*, ed. Stewart Sutherland and Peter Clarke (London: Routledge, 1988), 244.

15. See Nurit Bird-David, "Persons or Relatives? Animistic Scales of Practice and Imagination," in *Rethinking Personhood: Animism and Materiality*, ed. Miguel Astor-Aguilera and Graham Harvey (New York: Routledge, 2018), 25–34; Graham Harvey, "Bear Feasts in a Land without (Wild) Bears: Experiments in Creating Animist Rituals," *International Journal for the Study of New Religions* 9, no. 2 (2018): 195-214.

16. Lynn Margulis and René Fester, *Symbiosis as a Source of Evolutionary Innovation* (Cambridge: MIT Press, 1991).

17. Scott F. Gilbert, "When 'Personhood' Begins in the Embryo: Avoiding a Syllabus of Errors," *Birth Defects Research, Part C* 84, no. 2 (2008): 164–73; Scott F. Gilbert, "Holobiont by Birth: Multilineage Individuals as the Concretion of Cooperative Processes," *Arts of Living on a Damaged Planet*, ed. Anna Tsing, Heather Swanson, Elaine Gan, and Nils Bubant (Minneapolis: University of Minnesota Press, 2017), M73–M89.

18. Lynn Sagan, "On the Origin of Mitosing Cells," *Journal of Theoretical Biology* 14, no. 3 (1967): 255–74.

19. Graham Harvey, *Food, Sex and Strangers: Understanding Religion as Everyday Life* (New York: Routledge, 2013), 111–13.

BIRDSONG

David Taylor

The riot of birds in the wetlands
    as the sun rises above Swan Pond
tells me to notice the sun,
    rising over the dawn-etched pitch pines
and stretchy scarlet oaks.

My coffee is all the more sharp
because of the squawks of the bluejays,
    the northern flickers calling out,
    catbirds mimicking the remaining spring peepers,
    a Carolina wren flitting low across the yard to forest,
        nesting in the kayak left out over winter.

I guess it's the reason we all sing—partner, place, and
communal chorus,
    to have one's voice fit in with more than one,
        and in offering song have song returned.
    And now in morning, it's all the clearer.

When I listen intently to these morning offerings,
the words in the journal on the table
take wing too,
call out as though morning birdsong,
and lift from the page
when I read them aloud.
Hello. Good morning.

I'll keep writing, in song, hoping for words to take wing.

PERMISSIONS

These credits are listed in the order in which the relevant contributions appear in the book.

"When One Known to You Dies, the Rearranging of Space and Time Begins" was first published in *Kenyon Review* and appears here with the permission of Elizabeth Bradfield.

"Cornflowers" was originally published in Brenda Cárdenas, *Boomerang* (Bilingual Review Press, 2009) and appears here with permission.

"Charmed" first appeared in Susan Richardson, *Words the Turtle Taught Me* (Blaenau Ffestiniog, Wales: Cinnamon Press, 2018) and is reprinted here with permission.

"Dispatch from This Summer" was first published in *About Place* and appears here with the permission of Elizabeth Bradfield.

ACKNOWLEDGMENTS

As editors, we want to offer some deeply felt gratitude. An initial gathering in 2018—enabled by the generosity of the Center for Humans and Nature—began the conversations that would eventually become the *Kinship* series. During our time together, our group of twenty or so people often sat in a loose circle—some in chairs, some leaning against couches, some cross-legged on the floor—listening to unique experiences from all varieties of places. About this gathering, mostly it is the joy we remember, the laughter that seemed to bubble up spontaneously, reinforcing bonds of kinship and love for this living earth as we attempted to put this into words and actions.

We also recall the ways other voices came into our midst. During one of the meetings, for example, a participant in the middle of the circle pulled out her iPhone and gently asked us to listen. The sound of a family of Orcas communicating with one another entered the room. Even those of us who don't live anywhere near the Pacific Ocean felt deep recognition, made more poignant by the current threats faced by these fellow mammals. They were speaking to one another, yet it felt as though their voices were also reaching through the water to us. Kin.

Later, we ventured outside to stretch our legs and to participate in a soundwalk to attune ourselves to the nearby forest and its aural textures. We paused at a huge Oak whose sprawling crown seemed to cover most of the backyard. In a way, this long-lived Oak was the reason we gathered where we had. This tree's presence preceded the home in which we were meeting by many decades, likely centuries, and the home was there because the parents of

Strachan Donnelley, the founder of the Center for Humans and Nature, chose to live in this spot. The family story has it that this elder Oak tree drew them to the place. We paid our respects. Some of us laid a hand on the tree. All of us breathed, thinking about plant kin who outlive us in age and who gift us with oxygen.

On the basis of this meeting, the initial vision for *Kinship: Belonging in a World of Relations* was the creation of a single volume. Then it grew. As each of us, as editors, reached out to people of various expertise, asking them to share their perspectives and stories of kinship, new threads were suggested to us and the web became larger. Soon it was apparent that the web could not be contained by a single volume—at least one that abided by the rules of standard publishing specs. We decided to ask, What if? What if we let form follow function? What if we let this book become what it wants to become? What if this book should be a series? This line of questioning begat logistical problems. For one, publishers in an already-risky landscape can't comfortably take those kinds of chances.

Yet challenges can also create innovations. The *Kinship* series has thus become the Center for Humans and Nature's first venture into book publishing—of many, we hope. We're thrilled with the outcome. We'd like to express our deep gratitude for human kin that helped bring this ambitious project to such a beautiful result. Emily Lonigro, Demetrio Cardona-Maguigad, Felix Castellanos, Carla Levy, and Stacey Saunders of LimeRed, for the gorgeous book covers and visual design. Minds blown. Manuscript editor Katherine Faydash, for her expert eye and enthusiastic support of the content. Riley Brady, for the lovely page layout and design. Ronald Mocerino at the Graphic Arts Studio Inc., for meeting our every printing need. Chelsea Green Publishers, especially Michael Weaver and Michael Metivier, for being our fine confidants in distribution and promotion. Paul and Sandy Quinn, for hosting us so graciously at Windblown Hill. For their giant spirits, supportive presence, and care-filled work, the Center for Humans and Nature kinfolk: James Ballowe, Hannah Burnett, Anja Claus, Katherine Kassouf Cummings, Jon Daniels, Brooke Hecht, Bruce Jennings, Curt Meine, and Jeremy Ohmes; and for the support of the Center for Humans and Nature board: Gerald Adelmann, Julia Antonatos, Jake Berlin, Ceara Donnelley, Tagen Donnelley, Kim Elliman, Christopher Getman, Charles Lane, Thomas Lovejoy, Ed Miller, George Ranney, Bryan Rowley, and Eleanor Sterling.

Gavin would like to give extra thanks and love to his family—Marcy, Hawkins, and Peanut—for indulging him as a "nature nerd" and for keeping his heart full. Also, Coyote, Magpie, and the Crab-like Orbweaver—the world wouldn't be the same without you.

Robin offers gratitude to her human and more-than-human kinfolk for their loving support: Family, Students, Maples, Orioles, Foxes, Peepers, and the whole dazzling web of relations.

John would like to thank his wife, Karen, for her near quarter century of partnership in cultivating kinship, from raising our daughters Atalaya and Sol to sharing work as teachers in the School of Environment and Sustainability at Western Colorado University to building Camp Alpenglow in the heart of the Gunnison Country. None of John's work is possible without the mountains and snows that frame and stand sentinel above his valley, providing greater-than-human family since he first visited from New Jersey at age sixteen.

CONTRIBUTORS & KIN, VOLUME 4

Elizabeth Bradfield is the author of *Toward Antarctica, Once Removed, Approaching Ice,* and *Interpretive Work,* as well as *Theorem,* a collaboration with artist Antonia Contro. Her work has been published in *The New Yorker, Atlantic Monthly, Kenyon Review,* and her honors include the Audre Lorde Prize and a Stegner Fellowship. Editor-in-chief of Broadsided Press, she works as a marine naturalist/guide and teaches creative writing at Brandeis University. www.ebradfield.com

Brian Calvert is a writer and editor based in Southern California. His nonfiction work builds from his career as a journalist, including a foreign correspondent and explores the intersections of environmentalism, journalism, poetics, political philosophy, and social justice. He is the former editor of *High Country News,* a magazine covering issues related to the Western United States, and a former Ted Scripps Fellow in Environmental Journalism. He holds an MFA in creative writing (poetry) from Western Colorado University and a BA in English from the University of Northern Colorado.

Photo by Roberto "Bear" Guerra

Brenda Cárdenas's books and chapbooks include *Boomerang* (Bilingual Press); *Bread of the Earth/The Last Colors* with Roberto Harrison; *Achiote Seeds/Semillas de Achiote* with Cristina García, Emmy Pérez, and Gabriela Erandi Rico; and *From the Tongues of Brick and Stone* (Momotombo Press). She also co-edited *Resist Much/Obey Little: Inaugural Poems to the Resistance* (Spuyten Duyvil Press) and *Between the Heart and the Land: Latina Poets in the Midwest* (MARCH/Abrazo Press). Cárdenas has served as Milwaukee's Poet Laureate, co-taught the inaugural workshop for Letras Latina's Pintura:Palabra: A Project in Ekphrasis, and is Associate Professor of English at the University of Wisconsin-Milwaukee.

Photo by Jeanne Theoharis

Shannon Gibney is a writer, educator, activist, and the author of *See No Color* (Carolrhoda Lab, 2015) and *Dream Country* (Dutton, 2018), young adult novels that won Minnesota Book Awards in 2016 and 2019. Gibney is faculty in English at Minneapolis College, where she teaches writing. A Bush Artist and McKnight Writing Fellow, her new novel, *Botched*, explores themes of transracial adoption through speculative memoir (Dutton, 2022).

Photo by Kristine Heykants

Graham Harvey is professor of religious studies at The Open University, UK. His research largely concerns "the new animism," especially in the rituals and protocols through which Indigenous and other communities engage with the larger-than-human world. His publications include *Food, Sex and Strangers: Understanding Religion as Everyday Life* (2013), and *The Handbook of Contemporary Animism* (2013).

Lyanda Fern Lynn Haupt is a writer and naturalist based in the Pacific Northwest. She is the author of several award-winning books, including *Crow Planet: Essential Wisdom from the Urban Wilderness* and *Mozart's Starling*. Her most recent title is *Rooted: Life at the Crossroads of Science, Nature, and Spirit.*

Photo by Tom Furtwangler

John Hausdoerffer, jhausdoerffer.com, is author of *Catlin's Lament: Indians, Manifest Destiny, and the Ethics of Nature* as well as co-author and co-editor of *Wildness: Relations of People and Place* and *What Kind of Ancestor Do You Want to Be?* John is the Dean of the School of Environment & Sustainability at Western Colorado University and co-founder of Coldharbour Institute, the Center for Mountain Transitions, and the Resilience Studies Consortium. John serves as a Fellow and Senior Scholar for the Center for Humans and Nature.

Photo by Keith Carlsen Photography

Brooke Hecht, seeking the space to explore life's big questions, joined the Center for Humans and Nature in 2005 as a Research Associate. She has been the President of the Center since 2008. Whether through the Center's Questions for a Resilient Future program or other Center initiatives, her work explores what it means to be human and what our responsibilities are to each other and the whole community of life.

Liam Heneghan is a Dublin-born writer living temporarily in the U.S. Midwest.

Painting by Carla Hayden

Robin Wall Kimmerer is a mother, botanist, writer, and Distinguished Teaching Professor at the SUNY College of Environmental Science and Forestry in Syracuse, New York, and the founding Director of the Center for Native Peoples and the Environment. She is an enrolled member of the Citizen Potawatomi Nation and a student of the plant nations. Her writings include *Gathering Moss* and *Braiding Sweetgrass: Indigenous Wisdom, Scientific Knowledge and the Teachings of Plants*. As a writer and a scientist, her interests include not only restoration of ecological communities, but restoration of our relationships to land. She lives on an old farm in upstate New York, tending gardens domestic and wild.

Dr. Andy Letcher has doctorates in Ecology (from the University of Oxford) and the Study of Religion (King Alfred's College, Winchester). He is a Senior Lecturer at Schumacher College, Devon UK, where he is the Programme Lead for the MA Engaged Ecology. He is currently researching ritual and animistic usage of psychedelics by contemporary British Druids. He is the author of numerous papers on subjects as diverse as environmental protest, animism, fairies, the revival of the Heathen lyre, and the distribution of mammals across continents. He is the author of *Shroom: A Cultural History of the Magic Mushroom*. He plays English bagpipes, low whistle, and Dark Age lyre.

Photo by May Woods

Freya Mathews is Emeritus Professor of Environmental Philosophy at Latrobe University, Australia. She is the author of over a hundred books, articles, and essays in the area of ecological philosophy. Her 1991 book, *The Ecological Self*, will be re-released in 2021 in the Routledge Classics Series. In addition to her research activities, she co-manages a private conservation estate in northern Victoria. She is a fellow of the Australian Academy of the Humanities.

Daegan Miller is a critic, essayist, and author of the critically acclaimed *This Radical Land: A Natural History of American Dissent*. He lives with his family on top of a hill, and among the woods, in Western Massachusetts. Find out more about Daegan's work at daeganmiller.com.

Susan Richardson is a writer, performer, and educator whose fourth collection of poetry, *Words the Turtle Taught Me* (Cinnamon Press, 2018)—which emerged from her residency with the Marine Conservation Society—was shortlisted for the Ted Hughes Award. She is currently writer-in-residence with the British Animal Studies Network, facilitated by the University of Strathclyde. Susan has performed her work on BBC2, Radio 3, Radio 4 and at festivals from Hay to Adelaide. She is currently writing a work of creative nonfiction, due to be published in 2022.

Kim Ruffin fell in love with the outdoors climbing trees, playing on dirt mounds, riding bicycles in flip-flops, and jumping off home-made ramps. Her family's legacy of nature stewardship inspired her to become an educator (Associate Professor of English, Roosevelt University), author (*Black on Earth: African-American Eco-Literary Traditions*), and trained nature and forest therapy guide. A constant witness to the restorative powers of being outside, she seeks out activities that are "easy like Sunday morning." There's no place like Earth, and she's one proud Earthling.

David Taylor is an Assistant Professor of Environmental Humanities in the Sustainability Studies Program at Stony Brook University. His writing crosses disciplinary boundaries and genres. Poetry books include *Palm Up, Palm Down* (Wings Press, 2017), *The Log from The Sea of Cortez: A Poetry Series* (Wings Press, 2015), and *Praying Up the Sun* (Pecan Grove Press, 2008). Upcoming and recent scholarly publications include: *Paumanok Rising Again: Long Islanders Reflect on Justice and Climate Change* (SUNY Press, forthcoming), *An Island in the Stream: Ecocritical and Literary Responses to Cuban Environmental Culture* (Lexington Books, 2019), and *Sushi in Cortez: Essays from the Edge of Academia* (University of Utah Press, 2015).

Gavin Van Horn is the Creative Director and Executive Editor for the Center for Humans and Nature. His writing is tangled up in the ongoing conversation between humans, our nonhuman kin, and the animate landscape. He is the co-editor (with John Hausdoerffer) of *Wildness: Relations of People and Place*, and (with Dave Aftandilian)

City Creatures: Animal Encounters in the Chicago Wilderness, and the author of *The Way of Coyote: Shared Journeys in the Urban Wilds.*

Manon Voice is a native of Indianapolis, Indiana, and is a multi-hyphenate—poet and writer, spoken word artist and film producer, hip-hop emcee, educator, social justice advocate, community builder and practicing contemplative. She has performed on diverse stages across the country in the power of word and song and has taught and facilitated art and poetry workshops widely. Her poetry has appeared in *The Flying Island, The Indianapolis Review, The House Life Project: People + Property Series, Sidepiece Magazine, The World We Live(d) In* anthology, *The Indianapolis Anthology*, and more. In 2018, Manon received a nomination for the Pushcart Prize in Poetry. Manon is also a 2021 Editorial Fellow with the Center for Humans and Nature.

Dr. Andreas Weber is a biologist, philosopher, and writer. His work focuses on a re-evaluation of our understanding of the living. He proposes to view—and treat—all organisms as subjects and hence the biosphere as a meaning-creating and poetic reality. Andreas teaches at Berlin University of the Arts and is Adjunct Professor at the Indian Institute of Technology in Guwahati. He contributes to major German newspapers and magazines and has published more than a dozen books, most recently *Enlivenment: A Poetics for the Anthropocene* (MIT Press, 2019) and *Sharing Life: The Ecopolitics of Reciprocity* (Boell Foundation, 2020).

Photo by Florian Büttner

Brooke Williams writes about evolution, consciousness, and his own adventures exploring both the inner and outer wilderness. He believes that the length of the past equals the length of the future. He lives with the writer Terry Tempest Williams near Moab, Utah, where they watch the light and wait for rain.

Orrin Williams's work-life is as the Food Systems Coordinator for the Chicago Partnership for Health Promotion, a program of the Office of Community Engagement and Neighborhood Health Partnership at University of Illinois at Chicago. Otherwise, he is a student of spiritual matters, particularly those related to acknowledging our relationships with Oneness and all beings. Orrin does that through gardening, reading, research, writing, and as the executive director of the Center for Urban Transformation in Chicago and the Roots Watering Hole podcast series with co-host and soil scientist Dr. Akilah Martin.